Ion Beam Technology and Applications

Edited by Ozan Artun

Published in London, United Kingdom

Ion Beam Technology and Applications
http://dx.doi.org/10.5772/intechopen.111060
Edited by Ozan Artun

Contributors
Nicholas White, Divya Gupta, Usha Rani, Rahul Singhal, Sanjeev Aggarwal, Carlos A. Camacho-Olguin, Arturo García-Borquez, Carlos-Alberto González Rodríguez, Hector Cruz Mejía, Marco Solorio Ávila, Collins Nana Andoh, Ulyana Bliznyuk, Aleksandr Chernyaev, Sergei Zolotov, Felix Studenikin, Victoria Ipatova, Aleksandr Nikitchenko, Ozan Artun

First published in London, United Kingdom, 2023 by IntechOpen
IntechOpen is the global imprint of INTECHOPEN LIMITED, registered in England and Wales, registration number: 11086078, 5 Princes Gate Court, London, SW7 2QJ, United Kingdom

British Library Cataloguing-in-Publication Data
A catalogue record for this book is available from the British Library

Additional hard and PDF copies can be obtained from orders@intechopen.com

Ion Beam Technology and Applications
Edited by Ozan Artun
p. cm.
Print ISBN 978-1-83769-108-1
Online ISBN 978-1-83769-109-8
eBook (PDF) ISBN 978-1-83769-110-4

For EU product safety concerns:
IN TECH d.o.o., Prolaz Marije Krucifikse Kozulić 3, 51000 Rijeka, Croatia,
info@intechopen.com or visit our website at intechopen.com.

Meet the editor

Dr. Ozan Artun obtained both a bachelor's degree and a master's degree in Physics from Trakya University, Edirne, Turkey. He pursued a Ph.D. in Physics at Yildiz Technical University, Istanbul. After one year, he was appointed to Bülent Ecevit University as a research assistant and received his Ph.D. in Nuclear Physics in 2015. Dr. Artun also has a master's degree in Computer Engineering. In 2018, he became an associate professor. He is also a reviewer for many international journals. His research interests include particle accelerators, radioisotope production, nuclear batteries, software development, nuclear structure, nuclear fuel, astrophysics, ionizing radiation, and artificial intelligence.

Contents

Preface

Ion Beam Technology and Applications discusses important topics in the use of ion beams, which have a wide range of applications in medicine, science, and engineering.

To understand the basic physical processes of ion beams, this book begins by considering the theoretical framework for the interaction of ion beams with matter in "Introductory Chapter: Ion Beam Technology and Applications" before moving on to a discussion of the research on ion beam applications.

Chapter 2, "Ion Beam Application to Nuclear Material Damage Assessment", discusses applications of ion beam technology in the evaluation of nuclear material damage.

Chapter 3, "Surface Microstructure Changes Induced by Ion Beam Irradiation", uses mathematical models in the literature to explain the experimental evidence because the phenomenon is quite complex to understand. For this aim, the material surface sputtering induced by the irradiation is investigated and is given experimental evidence of microstructure changes induced by ion beam irradiation based on certain irradiation conditions.

Chapter 4, "Oblique Ar^+ Sputtered SiC Thin Films: Structural, Optical, and Electrical Properties", presents the investigation of structural, optical, and electrical behavior of as-deposited and argon (Ar) sputtered silicon carbide (SiC), which has a wide range of applications in industries and research fields, in ion incidence angle of 500 via experimental facility 200 kV Ion accelerator at Ion Beam Centre, KUK.

Chapter 5, "Electron Beam Processing of Biological Objects and Materials", investigates electron accelerators, including the physical properties of the electron beam, dose ranges, computer simulation of the electron irradiation method, and the depth dose non-uniformity in objects irradiated with accelerated electrons. Additionally, this chapter compares methods used for the irradiation of polymers, transplantology objects, medical items, pharmaceuticals, and foods.

Finally, Chapter 6 "DC Parallel Ribbon Ion Beams for High-Dose Processes", highlights an important step forward in the development of ion beam technology. The chapter mentions the advantages of ribbon beams of heavy ions over cylindrical ions and the development of a new magnetic configuration in ion beam technology.

I would like to thank the authors for their valuable contributions to the book. I am also grateful to IntechOpen Service Manager Ms. Elena Vracaric for her help and support throughout the publishing process.

Dr. Ozan Artun
Associate Professor,
Faculty of Science,
Department of Physics,
Division of Nuclear Physics,
Zonguldak Bülent Ecevit University,
Zonguldak, Turkey

Chapter 1

Introductory Chapter: Ion Beam Technology and Applications

Ozan Artun

1. Introduction

Ion beams are charged particle beams comprising ions accelerated in particle accelerators, and they have a wide variety of applications in science, medicine, and engineering since ion beam technology is a key tool that has found vital applications. In the present day, the applications of ion beams for scientific and commercial purposes are remarkable. Especially in nuclear medicine, the charged particles produced by particle accelerators are used to treat and diagnose cancer. For this aim, not only ion beam technology but also particle accelerator technology used in ion beam technology also stands out. It is obvious that ion beam technology has branched into many fields in science and technology. In addition to therapy, low- and high-energy ion beams effectively lead to new developments in space applications, materials science, and atomic and nuclear physics [1–8].

In ion beam applications, there are modern technologies including the use of energetic ion beams for different science fields. Fundamental areas of application include microelectronics, space, energy, physics, materials sciences, etc. Frankly, the underlying science in the interactions between the ions beam and atoms in matter should be well explained. Hence, the interactions of charged particles with matter are experimentally investigated, and the obtained measurements are compared with the specialized simulation and theoretical data. Though fundamental physical processes are well understood, the credibility of the obtained data explanations is restricted by insufficient physical data, the examination of the unevaluated experimental data, and many discrepancies beyond the error limits in data reported by the authors. Notably, two major data are needed to understand the profile of the investigated material, namely, stopping power and cross-section data. The cross-section depends on processes such as direct, elastic and nonelastic scatterings, and equilibrium and pre-equilibrium reaction processes [9, 10]. On the other hand, the stopping power gives data defining the slowing of the ion in the material such as elements, compounds, alloys, *etc*. For this aim, the charged particle interactions between ions and materials should be well defined [8, 11].

2. Theoretical framework for ion beam interactions with matter

The charged particles interacting with matter lose their energy through ionization and excitation of the atoms. Therefore, the stopping power of matter may be

IntechOpen

explained by the mean energy loss per unit path length (dT/dx). Then, mass stopping power for heavy particles is given by [11]:

$$\left(\frac{dT}{\rho dx}\right)_c = \left(\frac{dT_s}{\rho dx}\right)_c + \left(\frac{dT_h}{\rho dx}\right)_c \tag{1}$$

where c is collision interactions in which s and h represent soft and hard collisions. The soft collision term by Bethe can be written as:

$$\left(\frac{dT_s}{\rho dx}\right)_c = \frac{2Cm_0c^2z^2}{\beta^2}\left[ln\left(\frac{2m_0c^2\beta^2H}{I^2(1-\beta^2)}\right) - \beta^2\right] \tag{2}$$

Where $C \equiv \pi\left(\frac{N_AZ}{A}\right)r_0^2 = 0.150Z/A\ \text{cm}^2/\text{g}$ where $r_0 = e^2/m_0c^2 = 2.818 \times 10^{-15}$ is classical electron radius, m_0c^2 represents the rest-mass energy of an electron, and $\beta = v/c$. H means arbitrary energy boundary between soft and hard collisions.

The equation can be simplified as follows:

$$k \equiv \frac{2Cm_0c^2z^2}{\beta^2} = 0.1535\frac{Zz^2}{A\beta^2}\frac{MeV}{g/cm^2} \tag{3}$$

On the other hand, the hard-collision statement can be given by:

$$\left(\frac{dT_h}{\rho dx}\right)_c = k\left[ln\left(\frac{T'_{max}}{H}\right) - \beta^2\right] \tag{4}$$

where T'_{max} is written as:

$$T'_{max} \cong 2m_0c^2\left(\frac{\beta^2}{1-\beta^2}\right) = 1.022\left(\frac{\beta^2}{1-\beta^2}\right) MeV \tag{5}$$

Eq. (4) can be written as:

$$\left(\frac{dT_h}{\rho dx}\right)_c = k\left[ln\left(\frac{2m_0c^2\beta^2T'_{max}}{I^2(1-\beta^2)}\right) - 2\beta^2\right] \tag{6}$$

Then, the mass stopping power can be given as follows [11]:

$$\left(\frac{dT}{\rho dx}\right)_c = 2k\left[ln\left(\frac{2m_0c^2\beta^2}{(1-\beta^2)I}\right) - \beta^2\right] \tag{7}$$

$$\left(\frac{dE}{\rho dx}\right) = 0.3071\frac{Zz^2}{A\beta^2}\left[13.8373 + ln\left(\frac{\beta^2}{1-\beta^2}\right) - \beta^2 - ln\left(I\right) - \frac{\delta}{2}\right] \tag{8}$$

where δ is density effect correction based on constants of the medium a, X_1, X_0, and C. δ includes three situations as follows [12]:

(i) $\delta(X) = 4.6052X + a(X_1 - X)^m + C\ (X_0 < X < X_1)$

(ii) $\delta(X) = 4.6052X + C\ (X > X_1)$

$$\text{(iii)}\quad \delta(X) = \delta(X_0) \times 10^{2(X-X_0)}(X \leq X_0) \tag{9}$$

where X is $\left[\log\left(\beta/\sqrt{1-\beta^2}\right)\right]$ [11]. In Eq. (8), the term I represents the mean excitation potential of the medium for materials [13]. This term was obtained by the quantum mechanical approaches and the experimental data given by Paul and Schinner [14] as follows:

$$\begin{aligned}
&(i)\quad I \approx 19.0\, eV\ (Z = 1)\\
&(ii)\quad I \approx 11.2 + 11.7 \times Z\, eV\ (2 \leq Z \leq 13)\\
&(iii)\quad I \approx 52.8 + 8.71 \times Z\, eV\ (13 < Z)
\end{aligned} \tag{10}$$

The different correction terms can be added to Eq. (8) such as the shell correction terms $(-C/Z)$. In addition to elements, Eq. (8) may be written for the compound and mixtures based on the assumption of Bragg's rule [11, 15]:

$$W_a = \frac{N_a A_a}{\sum_b N_b A_b} \tag{11}$$

$$\left(\frac{dE}{\rho dx}\right)_{comp.} = \sum_a W_a \left(\frac{dE}{\rho dx}\right)_a \tag{12}$$

where W_a is given the weight fraction of element (including N_a atoms), and A means the atomic weight. Furthermore, the mass stopping power for the electrons and positrons can be written as:

$$\left(\frac{dT}{\rho dx}\right)_c = k\left[ln\left(\frac{\tau^2(\tau+2)}{2(I/m_0c^2)^2}\right) + F^{\mp}(\tau) - \delta - \frac{2C}{Z}\right] \tag{13}$$

where the term τ is T/m_0c^2 and C/Z for elements is shell correction term. $F^{\mp}(\tau)$ is written for the electrons and positrons as follows [11]:

$$F^-(\tau) \equiv 1 - \beta^2 + \frac{\tau^2/8 - (2\tau+1)\,ln\,2}{(\tau+1)^2} \tag{14}$$

$$F^+(\tau) \equiv 2\,ln\,2 - \frac{\beta^2}{12}\left\{23 + \frac{14}{\tau+2} + \frac{10}{(\tau+2)^2} + \frac{4}{(\tau+2)^3}\right\} \tag{15}$$

3. Conclusion

Considering both theoretical and experimental research and applications, this book will provide a broad perspective on developments in ion beam technology and the future of ion technology. Therefore, the book focuses on expanding and improving the application areas of ion beam technology.

Author details

Ozan Artun
Department of Physics, Faculty of Science, Zonguldak Bülent Ecevit University, Zonguldak, Turkey

*Address all correspondence to: ozanartun@beun.edu.tr; ozanartun@yahoo.com

References

[1] Artun O. Production of polonium-208, 209 and 210 for use in nuclear battery via particle accelerator. Applied Physics A. 2020;**126**:386

[2] Artun O. Investigation of production of medical 82Sr and 68Ge for 82Sr/82Rband 68Ge/68Ga generators via proton accelerator. Nuclear Science and Techniques. 2018;**29**:137

[3] Artun O. Estimation of the production of medical Ac-225 on thorium material via proton accelerator. Applied Radiation and Isotopes. 2017;**127**:166-172

[4] Artun O. Investigation of production of samarium-151and europium-152,154,155 via particle accelerator. Modern Physics Letters A. 2019;**34**(20):1950154

[5] Artun O. Investigation of the production of cobalt-60 via particle accelerator. Nuclear Technology & Radiation Protection. 2017;**32**(4):327-333

[6] Artun O. A new production route of Gadolinium-148 for use in radioisotope thermoelectric generators. Applied Radiation and Isotopes. 2022;**185**:110222

[7] Artun O. The investigation of the production of Ac-227, Ra-228, Th-228, and U-232 in thorium by particle accelerators for use in radioisotope power systems and nuclear batteries. Nuclear Instruments and Methods in Physics Research Section B: Beam Interactions with Materials and Atoms. 2022;**512**:12-20

[8] Artun O. Calculation of the mass stopping powers of medical, chemical, and industrial compounds and mixtures. Nuclear Technology and Radiation Protection. 2018;**33**(4):356-362

[9] International Atomic Energy Agency (IAEA). Development of a reference database for ion beam analysis. In: IAEA-44 TECDOC-1780. Vienna: International Atomic Energy Agency; 2015

[10] Koning A, Hilaire S, Goriely S. Talys Manual 1.96. 2021. Available from: https://tendl.web.psi.ch/tendl_2021/talys.html/ [Accessed: 26 August 2023]

[11] Attix FH. Introduction to Radiological Physics and Radiation Dosimetry. New York, USA: John Wiley & Sons, Inc.; 1986. p. 607

[12] Sternheimer RM et al. Density effect for the ionization loss of charged particles in various substances. Atom Data Nuclear Data Tables. 1984;**30**:261-271

[13] Nabipour JS, Sardari D, Danil GHC. Sensitivity of the Bragg peak curve to the average ionization potential of the stopping medium. Romanian Journal of Physics. 2009;**54**(3):321-330

[14] Paul H, Schinner A. Empirical stopping power tables for ions from $_{3}$Li to $_{18}$Ar and from 0.001 to 1000 MeV/nucleon in solids and gases. Atom Data Nuclear Data Tables. 2003;**85**:377-452

[15] Bragg WH, Kleeman R. On the particles of radium, and their loss of range in passing through various atoms and molecules. Philosophical Magazine. 1905;**10**:318

Chapter 2

Ion Beam Application to Nuclear Material Damage Assessment

Collins Nana Andoh

Abstract

With a substantial rise in world temperature as a result of uncontrolled greenhouse gasses released, particularly by industrialized countries, the need for efficient and economical energy has been increasing, necessitating the addition of the energy mix of one that will not harm the environment. This has pushed the nuclear sector to develop more advanced nuclear power plant systems. Quality materials capable of withstanding extreme radiation doses, high temperatures, and occasionally harsh corrosive conditions are required. This has resulted in the examination of the current materials already in use by earlier generations of the reactor system to determine their usefulness in advanced ones that would be subjected to even tougher environmental conditions than before and also design new materials that are of such strength and abilities. Since it can be accurately regulated and less time-consuming, ion beam technology has been extensively used in accelerator facilities and computer simulations to analyze the radiation deterioration of these materials. This chapter shed further insight into the many applications of ion beam technology in the evaluation of nuclear material damage. Apart from irradiation damage assessment, it also provided some understanding of numerous areas in the nuclear sector where ion beam technology is used.

Keywords: ion beam technology, radiation damage, nuclear materials assessment, radiation damage simulations, nuclear fusion

1. Introduction

1.1 Overview

The ever-increasing population worldwide has put a toll on the need to produce efficient energy at a very fast pace to meet the demand. It is for this reason that the United Nations (UN), in setting its agenda for sustainability for 2030, included Goal 7, which targets affordable, reliable, and clean energy [1]. Moreover, due to the debilitating effect of climate change, most countries are either moving towards nuclear energy or adding nuclear energy to their energy mix to prevent excessive greenhouse gases from the coal used in powering the country's thermal plants. And this is also geared towards the achievement of the Sustainable Development Goals (SDGs) 7 aforementioned by member countries [1]. According to the International Atomic Energy Agency's Power Reactor Information System (PRIS), as of March 24, 2023, [2] there were 57 nuclear reactors under construction worldwide. China ranked first with 19 units, followed by India, with eight reactors under construction at the time. In all, about 18 countries were constructing at least one reactor unit.

Moreover, as the demand keeps increasing, the need for reliable materials to sustain the needed energy without any incidents, as happened in the cases of Chernobyl, and Fukushima, is on the rise. This has led to the constant assessment of the existing materials and even the design of new, safer, or reliable materials. The assessment of the behavior of these materials, under different conditions and environments, is because the qualities of engineering materials have hindered the performance of power-generating devices since time immemorial. Several of the materials that were employed in the design either failed through corrosion, embrittlement, creep, radiation damage, or fracture, among others. This has led to the suspension or shutdown of some existing reactors.

The nuclear industry, since its early days, has employed several materials in different areas. These nuclear materials depended mostly on the type of reactor plant being designed. But generally, materials are needed for structural/cladding, moderators and reflectors, control, coolant, and shielding for a better and longer operating period for nuclear power plants. Some of the major materials that have been employed over the past years are aluminum, beryllium, magnesium, zirconium, stainless steel, carbon, graphite, boron, cadmium, hafnium, water, concrete, etc. Generally, these materials that are used in nuclear reactor design can be grouped or classified into four categories. Thus, metals, ceramics, polymers, and advanced materials such as semiconductors.

However, due to the harsh corrosive environmental conditions that exist in the nuclear reactor, coupled with the high radiation dose and high temperature, these materials at some point failed or experienced a lot of defects. The formation, distribution, and interaction of point defects (vacancies and interstitials) and their clusters, such as Frenkel pairs (vacancy-interstitial pairs), interstitial loops, voids, vacancy clusters, inert gas bubbles, and radiation-induced dislocation segments and networks were mostly associated with the nature of the defect in crystalline materials. And since no material can escape from such defects or damages, materials are normally subjected to ion beams of equal radiation doses about that of the nuclear reactor to assess their response. The materials are then modified based on these defects from the ion beams, and this is done over and over again to be able to determine the radiation tolerance which will help in advancing the technology and in the design of reactors even for a higher dose and with a prolonged reactor lifespan.

Most of the time these advanced reactors are only designed after the required material types are achieved. And since such reactor technology is not even at the prototype stage, there would not be any such reactor for testing materials except to utilize the likes of accelerators and some computational tools to bombard these materials and vary their compositions to be fit for application in the design of reactors of such technology. This is exactly the procedure that the current generation, Generation IV energy systems, fusion reactor systems, and even the previous generations of reactors are going through or went through. Currently, no fusion neutron irradiation facility exists for materials testing with a fusion spectrum irradiation and it is not easy studying the neutron irradiation effect even in the current reactor systems. Hence, ion beam experiments have therefore been researched by several researchers and it has been employed by most industry players to assess the irradiation effects on materials already employed in the fission spectrum [3, 4].

1.2 Ion beam technology

Ion beams are streams of energetically charged particles that are normally deployed and directed to the surfaces of target materials to interact with the material in various ways and modify the properties of such materials or dope with a thin film [5, 6]. Mostly the level of interaction of these beams depends on the energy and composition of the ions on one hand and the properties of the materials being bombarded on the other hand. The directed ions either remove the material (sputtering) or implant it into the material to change its atomic and molecular arrangements and thereby changing its properties [7]. This technology can be classified as one of the oldest in material design, but several new applications are emerging each passing day as a result of the increase in knowledge of the technology and the high demand for its usage by industry players. The evolution of the industrial regime however led to an increased interest in this new technology worldwide.

Although some powerful inventions that resulted in the design of the accelerator took place around the 1930s, the ion beam technology (IBT) was first used by the US space program in the late 1950 and early 1960s [5]. However, commercial use of the technology was in 1970 in the space field and then it led to the evolution of a superior semiconductor industry which has been instrumental in the materials design and manufacturing sector. Even with the introduction of microelectronics, the interest of researchers and scientists in this area has been phenomenal resulting in several sophisticated materials and designs. Not only has it been used in the area of material science and engineering, but advancement in technology has also led to new fields and new areas of application of the technology in the environment, medicine, biology, semiconductor manufacturing, and space explorations among others.

1.3 Scope of the chapter

This chapter of the book is intended to introduce ion beam technology and its applications, with an emphasis on their usage in nuclear radiation damage assessment to material scientists and engineers who have little or no knowledge in the area. This chapter will cover the application in the nuclear industry and dive into the application in the area of radiation damage assessment. Then, some computational approaches that utilize ion beams in radiation damage assessment will be addressed. Important milestones in the area of application of ion beams to radiation damage assessment will also be unveiled.

2. Applications of ion beam technology in the nuclear industry

In the nuclear industry, IBT has been used in a variety of applications. The technology has been used in sectors such as nuclear medicine, nuclear materials science and engineering, nuclear physics research, nuclear energy generation, and nuclear isotope development, among others. Nonetheless, these applications may be divided into four broad categories: ion beam analysis, isotope production, nuclear fission and fusion research, and ion implantation.

2.1 Ion beam analysis

Ion Beam Analysis is founded on the basic physics of interactions between incident particles and target atoms (IBA) [7, 8]. IBA has been used extensively in the nuclear sector using the four standard methodologies [9]. In the nuclear industry, IBA is usually used in determining the composition of nuclear fuels, undertaking radiation damage studies, characterization of nuclear waste, and investigating nuclear incidents and accidents [10]. Rutherford Backscattering Spectrometry (RBS), Elastic Recoil Detection Analysis (ERDA), Nuclear Reaction Analysis (NRA), and Proton-Induced X-ray Emission (PIXE) are the traditional techniques used. The RBS, NRA, and ERDA methods are categorized as nuclear methods, whereas the PIXE method is classified as atomic [7]. Particle-Induced Gamma Ray Emission (PIGE), Charged Particle Activation Analysis (CPAA), Scanning Transmission Ion Microscopy (STIM), Ionoluminescence, Secondary Ion Mass Spectroscopy (SIMS), and Ion Beam Induced Charge imaging (IBIC) are some of the novel methods available.

RBS is one of the nuclear techniques used extensively in the nuclear industry for the investigation of materials' near-surface layers. It takes its history from Sir Ernest Rutherford's scattering experiment in 1911 which led to the discovery of the atomic nucleus. The technique is primarily used to quantify the composition of a material and measure the elemental depth profiles [7]. The method's usual parameters are 1.5 to 14 μm H^+ or He^+ ions with energies ranging from 500 keV to 4 MeV. RBS records the elastic backscatter of the projectiles from the sample nuclei. Moreover, the measured energy is influenced by the mass of the target nucleus as well as the depth of the scattering event beneath the surface. It is non-destructive and very sensitive to heavy elements. Usually, an energy-sensitive detector—typically a solid-state detector—records the energy of the backscattered particles. The drawback of the technique is that it requires a combination of other techniques such as NRA and ERDA.

ERDA involves inducing elastic scattering in a material with ion beams. This leads to the determination of the yield and energy of particles ejected out of the surface region of samples under the bombardment [11]. Quantitative measurements from the experiment then help to determine the composition of the material. Elastic recoil detection analysis as was opined earlier normally comes in to complement the RBS. In RBS, the incident particles backscattered from target atoms are detected, whereas, in ERDA, the forward recoil target atoms are rather detected directly. The technique's application is mostly on depth profiling of multiple light elements.

Nuclear Reaction Analysis as the name suggests is where incident ion beams induce nuclear reactions (radioactivity) which are used to determine the elemental composition of materials. The technique works on the detection of radiation from the interaction of the projectile and nucleus. The NAR technique involves the study of the reactions of hydrogen at the surface and subsurface of materials. Nuclear activation analysis is also another type of NRA used by several facilities to determine the elemental composition of materials. NRA is a good method for measuring the concentration of certain elements (impurity or implanted ions) and their depth distribution in biological samples.

The PIXE technique involves the generation of x-rays as a result of bombarding the material under study with an ion beam [6]. This ion beam displaces electrons in the inner shell thereby forcing electrons in the outer shells to fill the space and in the process dissipates energies in the form of X-ray. The x-ray released characterizes the elemental composition of the material under study. This technique is also non-destructive, works faster, and is accurate.

2.2 Isotope production

One of the most important operations in the nuclear sector is the use of isotopic methods to create radioisotopes and radiopharmaceuticals (atomic energy in peacetime) [12, 13]. As a result, facilities like the Brookhaven Linear Isotope Producer, as well as others across the world, are being built to generate a range of radioisotopes and radiopharmaceuticals for delivery to nuclear medical and industrial communities for research and commercial uses. Additional facilities for isotope manufacturing include research nuclear reactors, accelerator facilities, and a variety of separation facilities. By hitting an element with a particle, radioactive isotopes can be created (a-particle, deuteron, proton, electron, neutron, and even high-energy x-rays). Particle beams of 100 s of MeV energy are typically employed at these facilities to produce these radioisotopes and radiopharmaceuticals for nuclear medicine, pharmaceutical industries, science, environmental agencies, and even industrial purposes.

There are over 1000 radioisotopes routinely generated in industrial facilities, but Technetium-99 m (TC-99 m) has attracted great attention due to its use in medical centers. Almost 10,000 hospitals worldwide today use these isotopes for cancer treatment and a variety of other nuclear medicine therapies [14]. Production is a daily task due to the pace of decay. Because of the enormous demand for these radioisotopes, there are often shortages or unstable supplies of Technetium-99 m (TC-99 m) for brachytherapy.

The International Atomic Energy Agency (IAEA) occasionally supports and instructs facilities involved in the manufacture of these radioisotopes, particularly research reactors that prefer adding isotope productions to supplement those generated by other isotope production facilities. In Ghana, for example, the Ghana Research Reactor 1 (GHARR-1) is not now capable of creating radionuclides, although it has the capability of doing so shortly.

However, Egypt has been utilizing the Russian 2Mw reactor and a Norwegian-built radioisotope production plant to produce the following radio, isotopes: Iodine-131, sodium-24, potassium-42, chromium-51, phosphorus-32, and colloidal gold-198 since the sixties.

2.3 Nuclear fusion and fission experiments

Ion Beam has been instrumental in the advancement of knowledge in the area of nuclear fusion and fission experiments. Most of the candidate materials that are studied over and over again for the implementation of these new nuclear reactor systems or Generations have been made possible due to innovative techniques employed under this technology. The Generation IV reactors which are to operate at high fluence and even the fusion reactor which is proposed to operate at a 14 MeV neutron source will be too difficult to study candidate materials under such conditions using test reactors [15, 16]. It is therefore good that technologies such as these are available so that accelerator facilities and computer systems can test these materials with ease. Even the cost and time involved in testing the candidate materials with test reactors in the fusion energy regime cannot be overemphasized.

Radiation damage – the production of collision cascades leading to Frenkel pairs formation (**Figure 1**) leading to the formation of point defects – is well understood through this technology [10]. The technology has helped to identify the levels of damage by different materials and this has even led to the classification of materials for different purposes. For instance, researchers have realized how strong tungsten is and hence recommended it used as the first wall of the fusion reactor system [16]. Even after such recommendations as a prime candidate for the first wall of such a reactor/device a lot of research and testing works are being carried out to affirm its ability to withstand the level of radiation that will be produced.

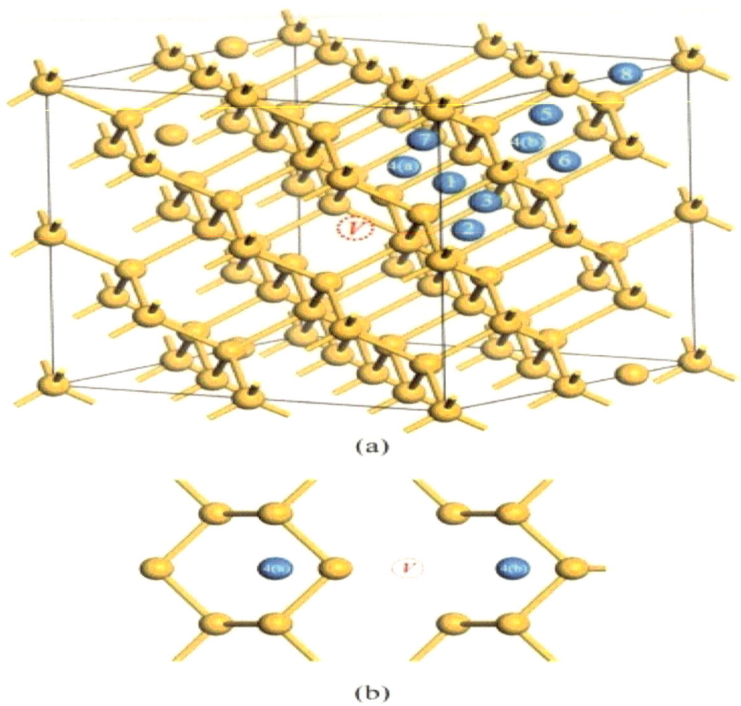

Figure 1.
Frenkel pairs in crystals of silicon adopted from [17].

Furthermore, the development of IBT, together with advances in knowledge, has resulted in a high pace of research originating from the fields of nuclear fusion and fission [16, 18]. The accelerator facilities have supported and armed researchers with the capability to investigate any material for use in a nuclear reactor without concern. It has relieved the strain of needing to raise criticality in nuclear reactors to test materials, among other things.

2.4 Ion implantation

The availability of IBT has also resulted in the creation of a variety of materials used in the nuclear sector. Ion implantation is the bombardment of accelerated ionized atoms (ion beams) or molecules to the surface of a material (target) to change its surface properties without changing their bulk properties [7, 19, 20] and render it suitable for use in various settings. This technique is generally carried out at a lower temperature. Ion implantation is used to develop new surface alloys and adjust surface-related attributes like hardness, toughness, fatigue, adhesion, wear, friction, dielectric properties, magnetic properties, and superconductivity as well as material corrosion resistance [21]. Ion beam coatings have also been shown to increase the mechanical, electrical, chemical, and tribological characteristics of materials used in industry. Improved material qualities are the foundation of better materials for production and design [22].

The use of this technology has been present for decades and continues to increase, particularly in the current period when practically every item is being reduced and so requires multiple microelectronic components. Long before the nuclear industry used ion implantation, the semiconductor industry was known to have employed this method to process semiconductor production components [23]. Nevertheless, the use of IBT in the nuclear sector has resulted in a plethora of unique research projects and methodologies, resulting in the industry's expansion in delivering novel materials for design purposes. Whereas the previous generation of nuclear power plants limited the use of electronic devices to an unavoidable extent, recent nuclear plants rely on electronics not only in digital computers and process control systems in a mild environment but also in harsh radiation conditions.

Several approaches to ion doping are employed by the industry. Some of the few ones are Ion Implantation (II), Ion Beam Mixing (IBM), Ion Beam Assisted Deposition (IBAD), Plasma Source Ion Implantation (PSII), Plasma Source Ion Deposition (PSID), Metal Vapor Vacuum Arc (MEVVA). These approaches have their pros and cons making them accepted in different spheres. All of these and many other approaches have been widely employed in the ion implantation of materials for industrial purposes. The approach chosen will be determined by the target material and the sort of testing being performed.

Another important use of Ion Beam Technology in the nuclear sector is radiation damage assessment. As a result, the next section will go over the topic in depth.

3. Ion beam technology applications in nuclear material damage assessment

As mentioned in the section on the employment of ion beams in nuclear fusion and fission experiments, IBT has proven a strong instrument in the evaluation of radiation damage. It's tempting to assume that the development of IBT has made studying radiation damage or irradiation processes conceivable. Because of the convenience and low cost of the IBT, several research activities and initiatives have been done in measuring nuclear radiation damage to nuclear materials. Ion beams are being used to assess material performance for nuclear energy applications at a rapid pace because ion irradiations produce radiation effects data that is directly relevant to understanding neutron-induced displacement damage [3]. Generally, the IBT is employed for:

1. Radiation Damage Characterization
2. Radiation hardness testing
3. Radiation damage mitigations
4. Dosimetry

3.1 Radiation damage characterization

Characterization of irradiation effects in materials using heavy or lesser ion beams is not new; it has been used for more than a decade. Tests with protons, for example, have gotten a lot of attention. Such experiments have been used as stand-ins for research into radiation in various nuclear materials. Yet, as the year passes and materials continue to fail under particular situations, the necessity to identify new materials to fit into such applications becomes more critical. Some programs and earlier studies have been successful in identifying materials suitable for use in nuclear reactor system design using the IBT.

Positron Annihilation Spectroscopy (PAS) experiments have also been used effectively to evaluate radiation damage in nuclear materials. Sabelová et al. [24] used positron annihilation techniques to analyze helium-implanted Fe-Cr alloys. Similarly, A review study by [25] stipulated and disclosed a lot of work on the use of positron annihilation in characterizing materials when it comes to radiation damage. The employment of a positron, according to the study by [25], is the only investigation that can precisely determine the size, concentration, and chemical makeup of individual atomic vacancies, as well as microscopic and large vacancy clusters formed by irradiation. The PAS was further supported by the study due to its extraordinary sensitivity to lattice defects. Doppler broadening spectroscopy, which refers to the widening of spectral lines induced by the Doppler effect, has also been utilized extensively together with PAS to evaluate materials.

Another approach used to characterize radiation damage in materials is a focused ion beam (FIB) [26–28]. It is comparable to Scanning Electron Microscopy (SEM), which has long been used in this capacity, but whereas SEM employs a focused beam of electrons to photograph the material in the chamber, a FIB apparatus utilizes a concentrated beam of ions. FIB may also be used in systems that include both electron and ion beam columns, enabling the same feature to be examined with either beam. It is one of the current applications of ion beams in characterization which is mostly deployed in the health sector but it is equally important in the spheres of nuclear materials damage assessment.

Transmission electron microscopy (TEM), atom probe tomography (APT), synchrotron radiation methods, micro-X-ray diffraction (XRD), and small-angle neutron scattering (SANS) are all potent techniques used to characterize radiation damage. All of these approaches use ion beam technology throughout the characterization process.

3.2 Radiation hardness testing

In the process of hardening components or materials for their use in the nuclear reactor especially where the nuclear flux is expected to be very high and with high-energy level photons, a procedure called radiation hardness testing is used. This procedure normally looks similar to the characterization of radiation damage as described in the previous section, however, in this case, the hardness of the material is the main objective for the test and nothing else. The material is subjected to a beam of ions to measure the response to that exposure. A certain elemental constituent is changed (either reduced in amount or increased) and then the new material is subjected to the same treatment again until the best material with the needed hardness is achieved.

Radiation hardness testing has been widely used in the spacecraft sector to ensure that materials for spacecraft can endure nuclear particle bombardment from both onboard and external sources [4]. Additional locations with radiation levels high enough to pose a risk to electronics are nuclear and high-energy physics experiments, as well as irradiation facilities in general, such as particle accelerators and nuclear reactors. Radiation hardness studies are particularly important for electronic instrumentation for patients' irradiation in cancer therapy.

It is critical to investigate a material's hardness before using it as a component in a reactor design, especially now that the reactor system is being shrunk to the point where semiconductors may be used. This is because the cost-effectiveness of such a technique cannot be compared to using a certain weak material and replacing it quickly due to its inability to withstand the reactor's specified life duration.

3.3 Radiation damage mitigations

Because of their capacity to create regulated levels of displacement damage under well-defined experimental settings, ion beam irradiations are chosen over neutron irradiations for many radiation effects research. Following that, ion beams aimed at a certain angle and with a regulated degree of displacement damage assist to rearrange the crystal lattice arrangement of these materials, giving them improved attributes appropriate for selection in nuclear reactor designs. This approach helps to mitigate the foreseeable damage of these materials when subjected to such harsh conditions.

As has been opined in the previous sections, ion beam technology is a major tool to mitigate the issue of radiation damage. This is through the process of assessing its suitability even before using it. Through the IBT, significant compressive surface stresses are produced which will partially compensate for externally imposed tensile stresses and lengthen component life against creep or fatigue failure by surface-initiated cracking.

In general, the significant information offered by radiation damage characterization and hardness testing level aids in the reduction of radiation damage in nuclear applications. It acts as a buffer against any unanticipated event or being caught off guard in any way. Moreover, it helps to decide on the dose of ion beams to introduce into the material to cause the required resistance.

3.4 Dosimetry

The rate at which materials absorb radiation is also critical in a high-radiation environment. As a result, ion beam technology is commonly used to examine how these nuclear materials absorb various types of radiation. The designer will then know where to put which material. Each component of the nuclear reactor system and its function; some absorb, some reflect, some moderate fast neutrons, while others prevent leakage. A material may qualify for leakage protection yet not be suitable for absorbing neutrons in the form of a control rod and vice versa.

Dosimetry is critical in advanced reactor designs, whether fission or fusion because the reactor systems are expected to sustain damage exceeding 30 dpa and to operate at temperatures ranging from 500 to 1000°C. And this type of damage and environment need tougher materials that will work properly and persist for the extended amount of time that modern reactors will be in operation. Even today's reactors, which have an average operational lifetime of roughly 40 years, are experiencing material wear and creep, let alone the expected 60 years [29].

Irradiation simulations are yet another key use of ion beam technology. Let us now have an overview of that.

4. Simulation techniques for ion beam technology application in radiation damage assessment

The nuclear business has also been significantly impacted by technological breakthroughs. The development of computers in the 20th century has been beneficial to nuclear scientists not only in the field of irradiation damage assessment but also in fields like thermal hydraulics, reactor control systems, and reactor operations, among others [30–34]. There are currently several computer programs and software packages available to conduct various nuclear industry simulations, particularly in damage assessments [30]. As a result of the computer's improved speed and memory capacity, carrying out such simulations is never a hassle for a researcher. The only thing needed from experimenters is an understanding of the ideas underlying the simulation and how the program functions.

Furthermore, damage evaluations performed using computer simulations are primarily due to the convenience and cost-effectiveness that experimenters find in simulating a condition on a computer rather than a testing facility. Additionally, with the aid of computer simulations, the atomistic level of the damage could be assessed easily as well. The good news is that in testing facilities, one may not be able to control every condition that could affect the assessment and must therefore rely on some assumptions. Yet, in the case of computer simulations or experiments, the researcher has control over the variables and so chooses the level of information necessary. For example, when modeling mechanical damage caused by radiation, COMSOL Multiphysics or ANSYS will provide all of the options. The size of the substance, the kind of radiation, and the other variables are all easily controllable. You may bombard the material with mixed radiations, add high temperatures, humidity, and other nuclear reactor-related variables, and run it all at the same time. Nevertheless, in the case of the accelerator, all of these requirements may not be present, thus certain assumptions must be made to account for this.

Many approaches, as illustrated in **Figure 2**, are available for modeling radiation damage at any level, whether atomic, molecular, or continuum. The various methodologies utilized in assessing material damage are for the continuum case Finite Element Methods (FEM) and Dislocation Dynamics (DD), and that of the atomistic

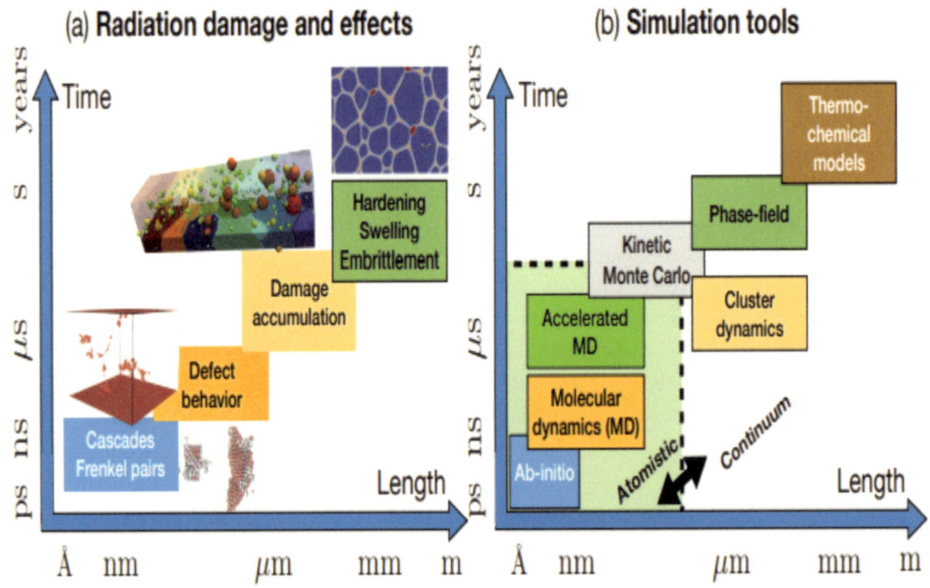

Figure 2.
Illustration of typical radiation damage simulations techniques across multiple lengths and timescales adopted from [33].

[33] and molecular levels as Ab-initio, Kinetic Monte Carlo (KMC), Monte Carlo (MC), Binary Collision Analysis (BCA), and Classical Molecular Dynamics (MD) simulations [4, 10, 32].

Ion beam technology has been used in some form or another in practically all of the techniques mentioned above. Apart from approaches for the continuum and dislocation states, which do not usually entail bombardment of the crystal lattice of these materials, the atomic and molecular dynamics phases have undergone considerable investigations that have extensively used the IBT. Although this chapter will not discuss in depth these methods, they have undergone considerable studies and research and hence the need to mention them in this chapter.

5. Important historical milestone of ion beam technology applications to nuclear radiation damage assessment

For researching radiation damage in materials, ion beam technology has been widely deployed. Some significant milestones so far in the development of ion beam technology applications for nuclear radiation damage assessments include:

a. **Initial experiments:** Researchers began utilizing ion beams to explore the effects of radiation damage on materials in the 1960s and 1970s. These early studies focused on simulating the effects of neutron irradiation on materials using ion beams [5].

b. **Defect formation:** Ion beam technology was utilized to explore the creation of flaws in materials during irradiation in the 1980s and 1990s. Ion beams were used by researchers to induce flaws in materials that were comparable to those formed by neutron irradiation.

c. **Effects of irradiation on thin films**: In the 1990s, ion beam technology was utilized to investigate the effects of irradiation on thin films. Researchers were able to induce flaws in thin films with ion beams and investigate the subsequent changes in their characteristics.

d. **Focused ion beam (FIB):** FIB technology was developed in the late 1990s and early 2000s. Researchers were able to induce extremely localized damage in materials using FIB technology, which was valuable for researching the impact of radiation damage on specific sections of material [26–28].

e. **Microscopy Methods:** In the 2000s and 2010s, ion beam technology was utilized in conjunction with microscopy techniques such as transmission electron microscopy (TEM) and scanning electron microscopy (SEM) to investigate the effects of radiation damage on nanoscale materials.

f. **High-energy ion beams:** In recent years, researchers have developed high-energy ion beam technology that allows them to explore the effects of radiation damage in materials at even greater energies and dosages.

g. **Detailed mechanism:** For the deployment of new, highly controlled in situ interrogation of materials during irradiation to provide dynamic and mechanistic data for model development, the Nuclear Science User Facilities recently submitted a roadmap for the development and improvement of current U.S. ion beam irradiation technologies in university and national laboratory settings [3].

h. **AI application to irradiation damage:** Machine learning is currently being employed in understanding the irradiation mechanism of high-entropy materials [35].

6. Summary of findings

Ion beam applications have been widely employed in the assessment of nuclear material damage. High-energy ion beams are used to simulate the effects of neutron irradiation on nuclear materials, and the ensuing damage is examined using various analytical techniques. Ion beams have significant benefits over typical neutron irradiation techniques in assessing nuclear material damage. When compared to neutron irradiation, ion beams may be accurately regulated and damage can be generated in a shorter amount of time.

Moreover, ion beams enable the investigation of particular damage processes, which is not always attainable and has led to a better understanding of the fundamental processes involved in neutron irradiation. Ion beam applications have been utilized to investigate a variety of materials, including metals, ceramics, and composites. The findings have improved our understanding of the underlying mechanisms involved in radiation damage and the impact of radiation on material characteristics. As a result, ion beam applications have proven to be invaluable tools in assessing nuclear material damage. The technology has allowed researchers to explore the effects of radiation on materials at a fundamental level, which is critical for the development of novel materials for use in nuclear reactors and other uses.

Author details

Collins Nana Andoh
Department of Nuclear Engineering, Graduate School of Nuclear and Allied Sciences, University of Ghana, Accra, Ghana

*Address all correspondence to: collinsnanaandoh@yahoo.com

References

[1] United Nations.. The 2030 Agenda and the Sustainable Development Goals: An opportunity for Latin America and the Caribbean (LC/G. 2681-P/Rev. 3), Santiago. 2018

[2] International Atomic Energy Agency. Power Reactor Information System (PRIS). 2023. Retrieved on the 25th of March, 2023, https://pris.iaea.org/pris/worldstatistics/underconstructionreactorsbycountry.aspx

[3] Heidrich B, Pimblott SM, Was GS, Zinkle SJ. Roadmap for the application of ion beam technologies to the challenges of nuclear energy technologies. Nuclear Instruments & Methods in Physics Research Section B-Beam Interactions with Materials and Atoms. 2019;**441**:41-45. DOI: 10.1016/j.nimb.2018.12.022

[4] Mei L, Guo X, Jin K. Characterization of mechanical property degradation of ion-irradiated materials. Frontiers in Materials. 2022;**9**:849209. DOI: 10.3389/fmats.2022.849209

[5] Cuomo JJ, Rossnagel SM, Kaufman HR. Handbook of Ion Beam Processing Technology: Principles, Deposition, Film Modification and Synthesis. United States of America: Noyes Publications; 1989

[6] Nastasi M, Mayer JW, Wang Y. Ion Beam Analysis: Fundamentals and Applications. United States: CRC Press; 2014

[7] Schmidt B, Wetzig K. Ion Beams in Materials Processing and Analysis. Vienna: Springer Science & Business Media; 2012

[8] Hellborg R, Whitlow HJ, Zhang Y. Ion Beams in Nanoscience and Technology. Heidelberg: Springer Science & Business Media; 2009

[9] Perea DB. Ion-Irradiation-Induced Damage in Nuclear Materials: Case Study of a-SiO2 and MgO. Switzerland: Springer; 2018

[10] Rubel M, Primetzhofer D, Petersson P, Charisopoulos S, Widdowson A. Accelerator techniques and nuclear data needs for ion beam analysis of wall materials in controlled fusion devices. EPJ Techniques and Instrumentation. 2023;**10**(1):3. DOI: 10.1140/epjti/s40485-023-00092-7

[11] Arnold Bik WM, Habraken FHPM. Elastic recoil detection. Reports on Progress in Physics. 1993;**56**:859

[12] Rath AK, Sahoo N. Particle Radiotherapy: Emerging Technology for Treatment of Cancer. India: Springer; 2016

[13] Fedosseev VN, Wendt K, Rothe S, Chrysalidis K, Goodacre TD, Marsh BD, et al. Ion beam production and study of radioactive isotopes with the laser ion source at ISOLDE. Journal of Physics G. 2017;**44**(8):084006. DOI: 10.1088/1361-6471/aa78e0

[14] The European Nuclear Society. Ensuring The Future – Outlook on Medical Radioisotopes and Uranium Supply Chains in Europe. ENS. 2022b https://www.euronuclear.org/news/ensuring-the-future-outlook-on-medical-radioisotopes-and-uranium-supply-chains-in-europe/

[15] Schiller PW. The 14 MeV neutron source for the development of fusion reactor materials. Fusion Engineering and Design. 1995;**30**(1-2):191-197. DOI: 10.1016/0920-3796(94)00408-y

[16] Knaster J, Moeslang A, Muroga T. Materials research for fusion. Nature Physics. 2016;**12**(5):424-434. DOI: 10.1038/nphys3735

[17] Sueoka K, Fukuda H. Theoretical study on Frenkel pair formation and recombination in single crystal silicon. Journal of Crystal Growth. 2019;**520**:1-10. DOI: 10.1016/j.jcrysgro.2019.05.014

[18] Horváth Á, Rachlew E. Nuclear power in the 21st century: Challenges and possibilities. AMBIO: A Journal of the Human Environment. 2016;**45**(S1):38-49. DOI: 10.1007/s13280-015-0732-y

[19] Wei B, Wu W, Wang J. In-situ TEM investigation of helium implantation in Ni-SiOC nanocomposites. Materials. 2023;**16**(4):1357. DOI: 10.3390/ma16041357

[20] Ahmad I, Ezema FI. Introductory Chapter: Ion Beam Applications. London, UK, London, UK: InTech EBooks; 2018. DOI: 10.5772/intechopen.78966

[21] International Atomic Energy Agency. Surface Modification of Materials by Ion Implantation for Industrial and Medical Application, Final Report of a Co-Ordinated Research Project. Austria: IAEA-TECDOC-1165; 2000

[22] Groover MP. Fundamentals of Modern Manufacturing: Materials, Processes and Systems. 7th ed. United States: Wiley; 2019

[23] Streade CA. Application ion implantation in tooling industry. Nuclear Instruments and Methods in Physics Research B. 1996;**113**:161

[24] Sabelová V, Krsjak V, Kuriplach J, Petriska M, Slugeň V, Veterníková J. Characterization of helium implanted Fe–Cr alloys by means of positron annihilation methods. Journal of Nuclear Materials. 2014;**450**(1-3):54-58. DOI: 10.1016/j.jnucmat.2013.10.039

[25] Selim F. Positron annihilation spectroscopy of defects in nuclear and irradiated materials- a review. Materials Characterization. 2021;**174**:110952. DOI: 10.1016/j.matchar.2021.110952

[26] Aitkaliyeva A, Madden JW, Miller B, Cole J. Implementation of focused ion beam (FIB) system in characterization of nuclear fuels and materials. Micron. 2014;**67**:65-73. DOI: 10.1016/j.micron.2014.06.010

[27] Sezen M. Focused Ion Beams (FIB) — Novel Methodologies and Recent Applications for Multidisciplinary Sciences. London, UK, London, UK: InTech EBooks; 2016. DOI: 10.5772/61634

[28] Ogihara S, Imafuku Y, Yamamoto R, Kogo Y. Application of FIB technique to introduction of a notch into a carbon fiber for direct measurement of fracture toughness. Journal of Physics. 2009;**191**:012009. DOI: 10.1088/1742-6596/191/1/012009

[29] What's the Lifespan for a Nuclear Reactor? Much Longer Than You Might Think. Energy.gov. https://www.energy.gov/ne/articles/whats-lifespan-nuclear-reactor-much-longer-you-might-think

[30] Nordlund K. Historical review of computer simulation of radiation effects in materials. Journal of Nuclear Materials. 2019;**520**:273-295. DOI: 10.1016/j.jnucmat.2019.04.028

[31] Chen S, Bernard D. Recommendation for computing neutron irradiation damage from evaluated nuclear data. Journal of Nuclear Materials. 2022;**562**:153610. DOI: 10.1016/j.jnucmat.2022.153610

[32] Wu Y, Song J, Zheng H, Sun G, Hao L, Long P, et al. CAD-based Monte Carlo program for integrated simulation

of nuclear system SuperMC. Annals of Nuclear Energy. 2015;**82**:161-168. DOI: 10.1016/j.anucene.2014.08.058

[33] Deo C, Chen EY, Dingreville R. Atomistic Modeling of radiation damage in crystalline materials. Modelling and Simulation in Materials Science and Engineering. 2021;**30**(2):023001. DOI: 10.1088/1361-651x/ac2f83

[34] Andoh CN, Ayensu A, Banini GK. Molecular dynamics simulation of mechanical deformation of austenitic stainless steels (Fe-Ni-Cr alloys) at supercritical water conditions. Journal of Applied Science and Technology. 2017;**22**(1-2):1-13

[35] Zhao S. Application of machine learning in understanding the irradiation damage mechanism of high-entropy materials. Journal of Nuclear Materials. 2021;**559**:153462. DOI: 10.1016/j.jnucmat.2021.153462

Chapter 3

Surface Microstructure Changes Induced by Ion Beam Irradiation

Carlos A. Camacho Olguín, Arturo García Bórquez, Carlos A. González Rodríguez, Héctor Cruz Mejía and Marco Solorio Ávila

Abstract

As a study result of irradiation-induced damage, various mathematical models have been developed to explain the phenomenon of irradiation-induced surface sputtering, these models are currently used in technological fields as diverse as materials characterization and in thin film deposition used in the manufacture of electronic or medical components. However, the phenomenon is not fully understood due to its high complexity, this work has the goal to present experimental evidence of the changes induced in the surface of hypereutectic alloy Ni-22% at Si promoted by the nickel ion beam irradiation, the surface changes induced will be explained in terms of the Sigmund theory of surface sputtering.

Keywords: ion beam irradiation, sputtering yield, surface microstructure, preferential sputtering, chemical and geometrical sputtering

1. Introduction

Material irradiation, material bombardment with energetic particles in the electro-volt to mega-electro-volt range, generally promotes changes in the material microstructure as a result of the energy transference between the solid atoms and the particle used to bombing the solid. The induced microstructural changes by ion beam irradiation that have been investigated and reported during the past five decades are: surface erosion, cratering, atoms implantation at controlled depth, crystal lattice disorder, phase transformations, second phases precipitation, etcetera [1, 2].

Because ion irradiation can deposit high energy densities locally in relatively short time periods, it could be used to evade thermodynamic constraints and create microstructures containing non-thermodynamic equilibrium phases [3, 4].

- When a surface is bombarded by an energetic ion beam a rich variety of surface structures are developed at atomic, microscopic, and macroscopic scales. These structures include elevations, with respect to the surrounding surface, e.g., plateaus, cones, and pyramids among others, depressions, e.g., pores and cavities. These features may occur singly or in large numbers and often form repetitive patterns.

- The irradiation process induces the mass transport phenomenon as a result of the point defect formation and their posterior annihilation in point defect sinks [5]. In binary alloys characterized by solute-defect complexes formation with strong interactions, for example, Cu-Be and Ni-Si, a few point defects are required to transport a solute atom considerable distance, and dramatic changes in solute distribution would occur [6].

- It is known that the irradiation promoted the second phase's formation under conditions of temperature and solute concentration that thermodynamically do not correspond to their stability field. The non-equilibrium second phase formation is promoted by irradiation-induced segregation, this irradiation-induced segregation occurs at a temperature where atoms transport via irradiation-induced point defects is possible, but redistribution of solute atoms via thermal vacancies is not the dominant process. Experimentally has been determined that if the solute atoms are smaller than the solvent atoms, the solute atoms will migrate toward sinks (surfaces, grain boundaries, dislocations, etc.), and larger atoms decrease their concentration near these sinks [7].

The knowledge acquired about the basic mechanisms of the phenomena that occur during the interaction between the ion beam with the material used as a target, the knowledge acquired as a result of the damage induced by the radiation has permitted the development of new technology to manufacture materials with specific properties due at the high control that could have over the material structure [8].

The most spectacular example of the use of materials irradiation is the analysis technique known as FIB (Focused Ion Beam). The ability of this technique to grind, image, and deposit material is critically dependent on the nature of the particle-solid beam interaction.

1.1 Alloys sputtering

The material surface sputtering induced by the irradiation is explained by Sigmund's theory or with modifications of itself. The phenomenological model developed by Sigmund employs the collision cascade as the fundamental event to explain the material surface sputtering: the incident ion upon collision with the solid initiates a series of atomic collisions with the surrounding atoms in the interaction volume, these atomic collisions are denominated the collision cascade. The incident ion energy is shared among those atoms that are located within of collisions cascade and promoted atomic detachment and atomic mobility. Only those collisions that occur near of material surface are effective in ejecting atoms out of the material surface. The great majority of atoms removed come from the first atomic layers of the material surface. The sputtering yield is the average number of atoms ejected from the material surface per incident ion. The sputtering yield depends on several things: the incidence angle at which ions collide with the material surface, the incoming ions energy, ions/surface atoms mass ratio, and the material surface binding energy. At a higher number of atomic collisions that occur in a near-surface region, the surface Sputtering Yield would be higher.

The Sputtering Yields' main characteristics for complex materials as binary alloys are the same as a pure element, but there are additional complications since there are two different types of atoms in the material. The two species cannot be eroded at an equal rate due to differences in energy partitioning, being removed probability or binding energies. Experimentally, preferential sputtering of one atomic species over the other has been observed in many alloys or compounds [9].

1.2 Preferential sputtering

In the sputtering yield description of a multicomponent system, the preferential erosion influence and surface segregation must be included. For a homogeneous specimen with two components A and B, the surface concentrations, Ns, are equal to the volume concentrations, Nv, in the absence of segregation to the surface, which must occur due to thermal processes. At the sputtering onset.

$$N^{s}_{A} / N^{s}_{B} = N^{v}_{A} / N^{v}_{B} \tag{1}$$

The partial Sputtering Yield of the atomic species A and B is defined by: $Y_{A,\,B}$ = (atoms eliminated amount A, B)/incident particles.

The sputtering yield of species A, Y_A, is proportional to the surface concentration, N^s_A, and similarly, Y_B is proportional to N^s_B.

The partial sputtering yield ratio is given by:

$$Y_A / Y_B = r\left(N^{s}_{A} / N^{s}_{B}\right) \tag{2}$$

Where the erosion factor, r, considers the differences in surface binding energies, escape depths of the ejected atoms and energy transfer within the cascade. Measured values of, r, are generally in the range 0.5–2.

On the other hand, there are several mathematical models that describe the sputtering Yield, it is important to mention that all of these models are based on the work of P. Sigmund. His work "Theory of Sputtering I. Sputter yield of Amorphous and Polycrystalline targets" published in Physical Review is a benchmark in this field [10].

1.3 Sigmund mechanism

The sputtering process nature is discontinuous due to each one ion initiating an atomic collision cascade. The sputtering yield is dependent on the deposited energy fraction at the surface, which is the function of surface topography local and ion beam incidence angle. From atomic level point of view, the surface sputtering has a random nature as a result the sputtering yield varies markedly from one event to another. These two phenomena have important consequences with respect to the surface topography when considering dimensions similar to those of atomic collision cascade.

A particularly violent event can produce a crater by removing a large number of atoms, either by erosion or by sublimation from a hot spot. The sputtering yield variation with position leads to the prediction that in the absence of a surface smoothing mechanism such as diffusion; flat surfaces are unstable, against increasing roughness at scales similar to the atomic collision cascade size. This instability can lead to the growth of small irregularities of atomic dimensions, such as surface protuberances.

2. Experimental evidence of microstructure changes induced by ion beam irradiation

In order to explain the changes induced by irradiation on a complex surface formed by the presence of two regions with different chemical compositions, the Ni-22%at.Si (Ni-22Si) concentrated alloy was selected. The Ni-22Si alloy is of the hypereutectic type, consisting of two regions, the dendritic and the interdendritic, the dendritic region is constituted by the $Ni_3Si_{-monoclinic}$ phase, and in the interdendritic region there is a regular eutectic formed by the regular alternation of the Ni(Si) and $Ni_3Si_{-monoclinic}$ phases lamellae (**Figure 1**), it was determinate by optical microscopy that 70% of the surface is occupied by the $Ni_3Si_{-monoclinic}$ phase. The terminal Ni(Si) phase has a maximum solubility of 12 atomic % of Silicon at the irradiation temperature, on the other hand, the $Ni_3Si_{-monoclinic}$ phase is an intermetallic phase with a Silicon concentration of 25 atomic %.

In summary, the main characteristics of the hypereutectic microstructure are the presence of two crystalline phases with different chemical compositions and crystalline structures, as well as the high density of grain boundaries.

Before being irradiated, the surface was mirror-polished to eliminate protrusions or surface defects that could locally modify the sputtering yield. The surface was analyzed using atomic force microscopy (AFM) to establish the surface topography parameters, an AFM image of the Ni-22Si alloy surface before being irradiated is shown in **Figure 2**.

As a result of the MFA images analysis of the surface before being irradiated, the protrusions' existence with heights of angstroms has been detected, as well as the presence of a step between the dendritic and interdendritic regions, the dendritic

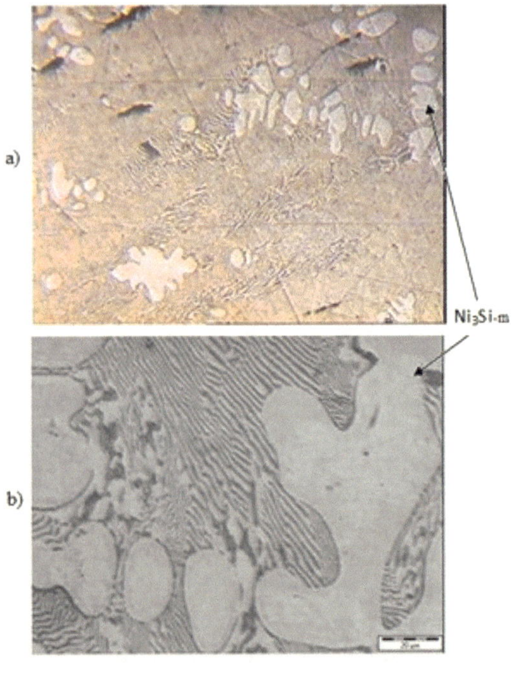

Figure 1.
Microstructure of the hypereutectic alloy (a) without chemical attack and (b) with chemical attack.

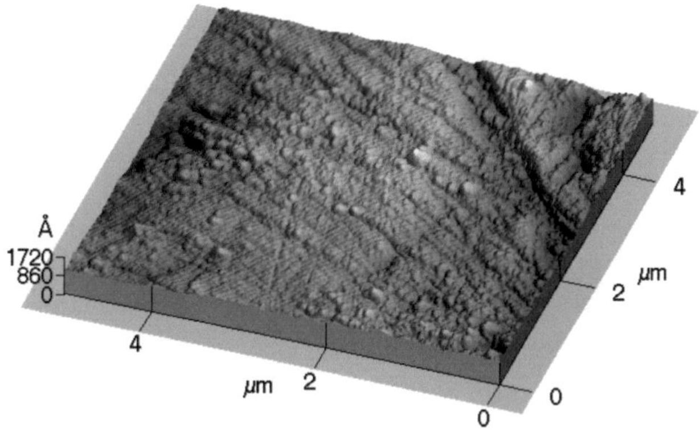

Figure 2.
Atomic force microscopy image of the Ni-22Si alloy surface before to be irradiated.

region is located on the upper right side of **Figure 2**, the interdendritic region occupies the rest of the surface shown in **Figure 2**.

2.1 Irradiation conditions

The Ni22at%Si was prepared by solidification in an inert atmosphere furnace, the alloy was templated from melting temperature to room temperature in water. The irradiation targets were obtained by electroerosion with the next geometrical characteristics: alloy circular plates with an area of 1 cm^2 and thickness of 300 μm. Prior to being irradiated the plate's thickness was reduced to 100 μm to a mirror finish by mechanical polishing. The irradiation conditions used to bombard the alloy circular plates were Ni ions beam of 3.66 MeV at 650° C, the incidence angle of the ion beam to the surface was 90°, two different ion fluence of 3.97×10^{16} and 1.99×10^{17} ions/cm^2 were used to evaluate the irradiation time effect. It was used a pyrometer focused on the irradiated area, with a tolerance of ±10°C to perform the temperature control of the irradiated area. The irradiation was performed in a Tandetron-type GIC linear accelerator, designed to provide a large flow of ions 1.9×10^{17} ions/cm^2 [11].

3. Analysis of the surface microstructure changes induced by the ion irradiation

Figure 3 shows, an optical microscopy image of the irradiated and non-irradiated regions of the Ni-22Si alloy, two regions are observed, one in bright contrast and the other one in dark contrast, the dark contrast region corresponds to the irradiated region. This change from bright to dark contrast is generated by the surface topography modification induced by irradiation, as a consequence, the reflective power of the irradiated surface is reduced. The surface topographic change is generated by the preferential sputtering of boundaries between the dendritic lobes and the interdendritic region, as well as by the crater's formation, dark spots, both changes are surface damage induced by irradiation [12].

As a result of several reported works on irradiation-induced damage in metallic alloys, it has been established that preferential sputtering of grain boundaries occurs because the atoms located in that region have lower binding energy with their neighbors [13].

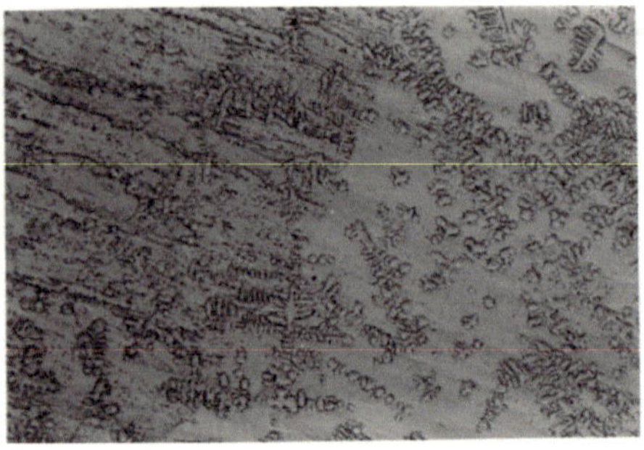

Figure 3.
Optical microscopy image of the irradiated and non-irradiated region.

3.1 Preferential erosion

Figure 4 shows an optical microscopy image of the boundary formed between the dendritic lobes and the interdendritic region, this image shows in detail the preferential sputtering suffered by the border between the two regions, the sputtering yield suffered by the border is greater than the erosion suffered by the dendritic region or the interdendritic region. In the image in **Figure 4**, it can also be seen that the sputtering of the boundary is not homogeneous, the alternation of protrusions and hollows is clearly observed, for this reason, atomic force microscopy was used to analyze in greater detail the sputtering process of grain frontier.

Atomic force microscopy images (**Figure 5**) of the irradiated region provide experimental evidence of sputtering process inhomogeneity, in **Figure 5(a)**, on the left side, a dendrite nodule (monophasic region) and on the right side of the same figure, the inter-dendritic region (biphasic region), and the border formed between the two regions can be seen. In **Figure 5(a)**, it is clearly seen that some specific regions of the boundary had suffered a greater sputtering giving rise to the bridges formation with a regular alternation, **Figure 5(b)** shows the same boundary at higher resolution, this last image allows us to establish that the sputtering process is indeed a non-homogeneous process since certain portions of the boundary were totally eliminated.

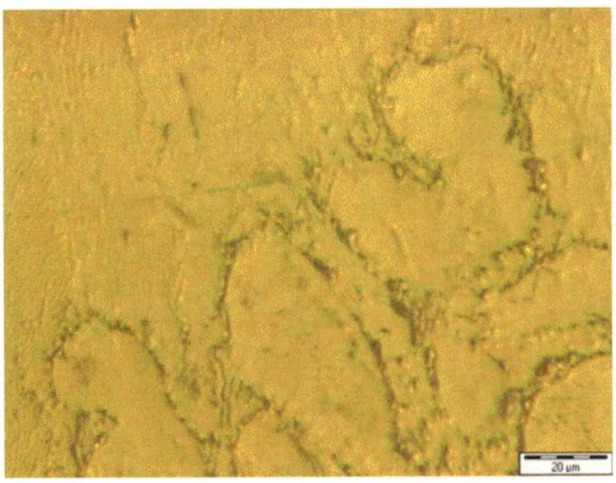

Figure 4.
Optical microscopy image of irradiated regíón.

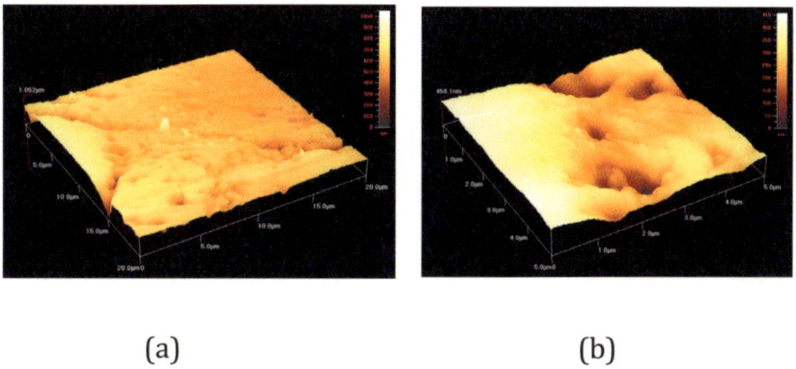

Figure 5.
Atomic force microscopy images of the boundary between dendritic and interdendritic regions.

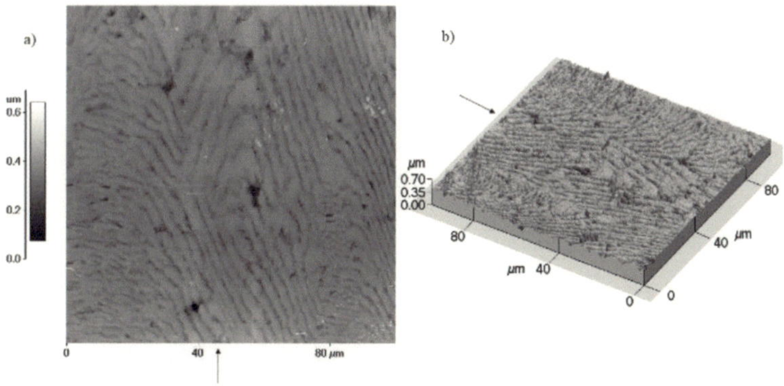

Figure 6.
Atomic force microscopy images of the valley-ridges pattern formed in the interdendritic region because of sputtering.

Figure 6 shows atomic force microscopy images of the interdendritic region to establish the origin of the sputtering inhomogeneity. Experimental evidence provided by MFA images (**Figure 6**) of the interdendritic region shows that the valleys and ridges regular pattern was formed as a sputtering result, this pattern regularity is very similar to the phases alternation that forms the Ni(Si)-$Ni_3Si_{\text{-monoclinic}}$ regular eutectic. Therefore, it is hypothesized that one of the phases was preferentially sputtering and that this preferential sputtering is promoted by the chemical difference in composition between the two phases.

The sputtering yield of each phase that forms the Ni(Si)-$Ni_3Si_{\text{-monoclinic}}$ regular eutectic was calculated using SRIM 2013 code [14]. **Table 1** shows the results of the sputtering yield simulation at the above-mentioned irradiation conditions.

Simulation results indicate that the terminal phase Ni(Si) has the highest sputtering yield. This means that this phase is the one that underwent the greatest sputtering under the irradiation conditions employed and consequently gives rise to the valleys formation observed in the atomic force microscopy images analyzed.

In addition, scanning electron microscopy images of the interdendritic region, **Figure 7**, show the same valley-ridges pattern, it should be remembered that the field depth of this microscopy is limited, however, the pattern characteristics are clearly observed.

Phase	Sputtering yield	Depth of maximum damage (μm)	Ion range (μm)
Ni(Si)	1.03	1.2	1.9
$Ni_3Si_{\text{-monoclinic}}$	0.81	1.3	2.0

Table 1.
Sputtering yield simulation by SRIM 2013 code [13].

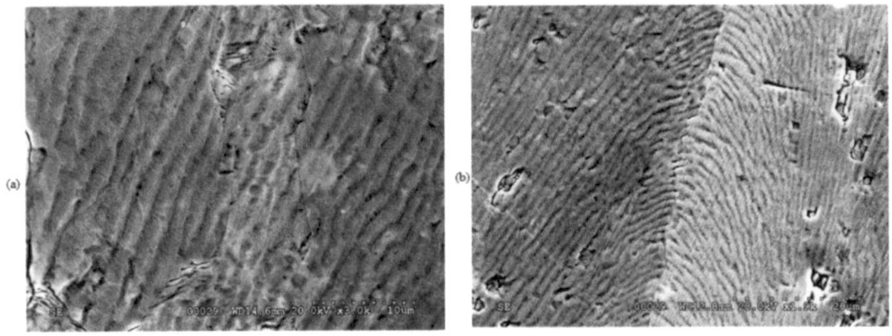

(a) (b)

Figure 7.
Scanning electron microscopy images of the interdendritic region irradiated, (a) interdendritic region and (b) eutectic colonies.

In the images shown in **Figure 7**, in addition to the pattern of valleys and ridges, other changes generated by irradiation can be clearly seen, craters and in **Figure 7(b)** it is observed how the boundary between eutectic colonies was sputtered, which denotes the high susceptibility to the chemical composition local variation of sputtering process induced by ion beam irradiation.

3.2 Surface protrusions evolution under ion irradiation

Figure 2 shows an MFA image of the surface before being irradiated in which it is observed that there are protuberances with angstroms heights without a defined pattern in its surface distribution, it has been established that such protrusions as an irradiation consequence result in the peaks formation with nanometer dimensions, which can be seen in **Figure 8**.

The roughness enhancement mechanism proposed by Sigmund [15] explains the formation of these nanostructures in terms sputtering yield variation generated as a consequence of the local incidence angle change of the ion beam.

In the case of three-dimensional, cone-like protrusions, the sputtering yield is about 0.2 in the cone zones with dimensions similar to or larger than the displacement cascades size. For protrusions with dimensions of the order of displacement cascade size the sputtering yield has values < 1, although this model predicts the nanoparticle's formation from protrusions with angstrom dimensions, its major contribution is that it predicts the most important geometrical feature, the nanoparticles top will always tend to have a conical shape. As an above result and in the absence of any surface smoothing mechanism by migration, pre-existing surface protrusions on the

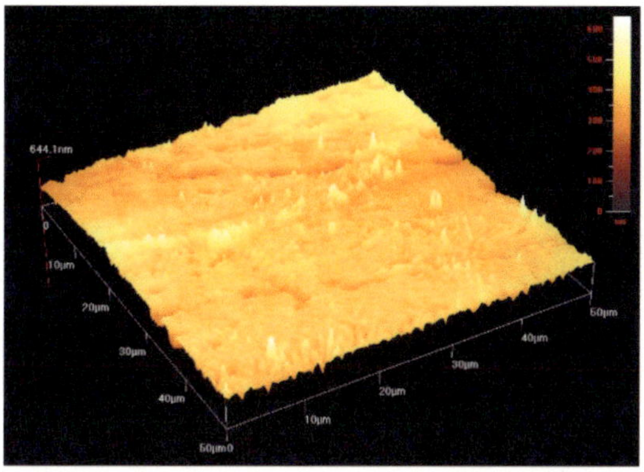

Figure 8.
Atomic force microscopy image shows a panoramic view of the interdendritic region of the irradiated zone where the distribution and size of the induced nanoparticles can be observed as a result of the interaction process of the irradiation with the protrusions observed in **Figure 2**.

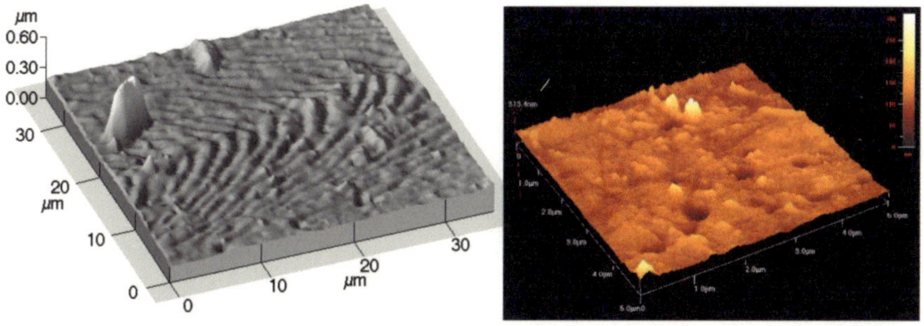

(a) (b)

Figure 9.
Nanopicks formed in the (a) interdendritic zone and in the (b) dendrites branches.

irradiated surface will be stable and will tend to grow as a result of the sputtering yield variation promoted by the local incidence angle change.

Figure 9(a) and (b) shows MFA images of the dendritic (**Figure 9b**) and interdendritic (**Figure 9a**) regions of the irradiation-affected area, both figures show the formation of peaks, cones, and other types of geometries and all of them have peaks that tend to have a conical shape. Another important factor detected is that the sputtering originated by chemical composition is of lower intensity than the geometrical sputtering, induced by incidence angle local variations of the ion beam.

3.3 Microstructural elements effect over the sputtering yield

When analyzing the changes generated by irradiation with nickel ions in the hypereutectic microstructure of the Ni-22Si alloy (**Figure 2**), the following results were obtained, it was detected that the dendrites ($Ni_3Si_{\text{-monoclinic}}$) have a lower sputtering yield with respect to the interdendritic region.

Figure 10(a) and **(b)** shows the sputtered surface profile of two dendrites because of ion beam irradiation.

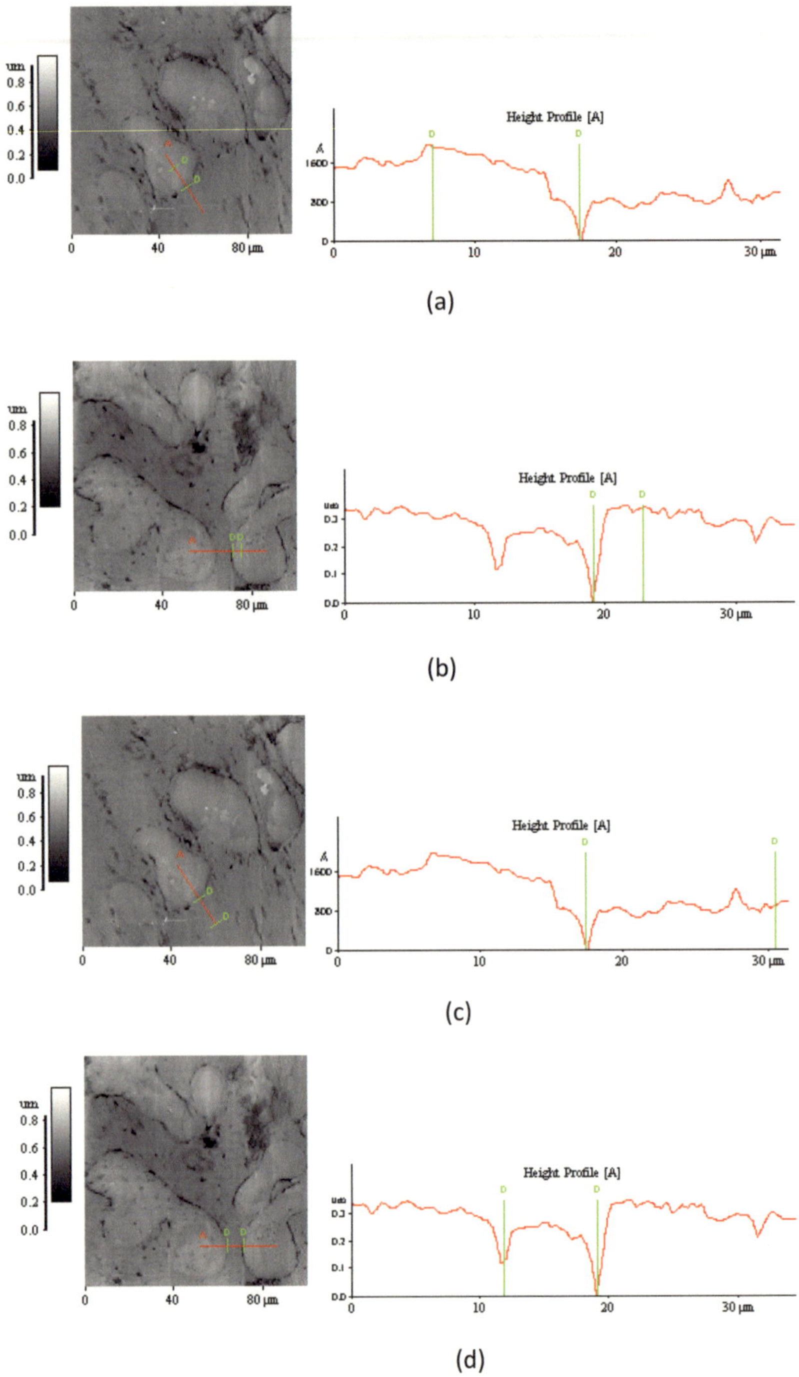

Figure 10.
Atomic force microscopy images and height profiles of the microstructural elements shown.

The regions delimited by the vertical lines are part of the dendrite(s) seen in the image to the left of the height profile, it is clearly determined that the dendrites have a lower atomic erosion ratio with respect to the interdendritic region.

Figure 10(c) presents the height profile of the interdendritic region shown in **Figure 10(a)** and clearly shows that on average the interdendritic region has a lower height with respect to the attached dendritic region, the initial surface before being irradiated did not have this type of height variations among its microstructural constituents.

On the other hand, **Figure 10(d)** shows the interdendritic region between the two dendrites shown in **Figure 10(b)**. Once again it is clearly seen that the interdendritic region has a lower height with respect to the height of both dendrites. On the other hand, there is another microstructural element in these images that must be analyzed; this is the grain boundaries and interfaces.

In conclusion, it can establish that the monophasic region, i.e., the dendrites, has a lower sputtering yield compared to the biphasic region, the interdendritic region and that because of this difference in sputtering yield, the difference in level between the dendrites and the interdendritic region is formed.

3.4 Effect of grain boundaries and interfaces on the sputtering yield

In order to analyze the grain boundaries effect on the sputtering yield, the hyper-eutectic alloy was irradiated with two doses, 80 and 380 dpa, and the grain boundaries effect was evaluated using Force Atomic Microscopy. **Figure 11** shows AFM images of the irradiated region with both regions, the height profile, and the correspondent image of the surface irradiated at both doses, the image at 80 dpa is shown in **Figure 11(a)** and the image at 380 dpa shown in **Figure 11(b)**.

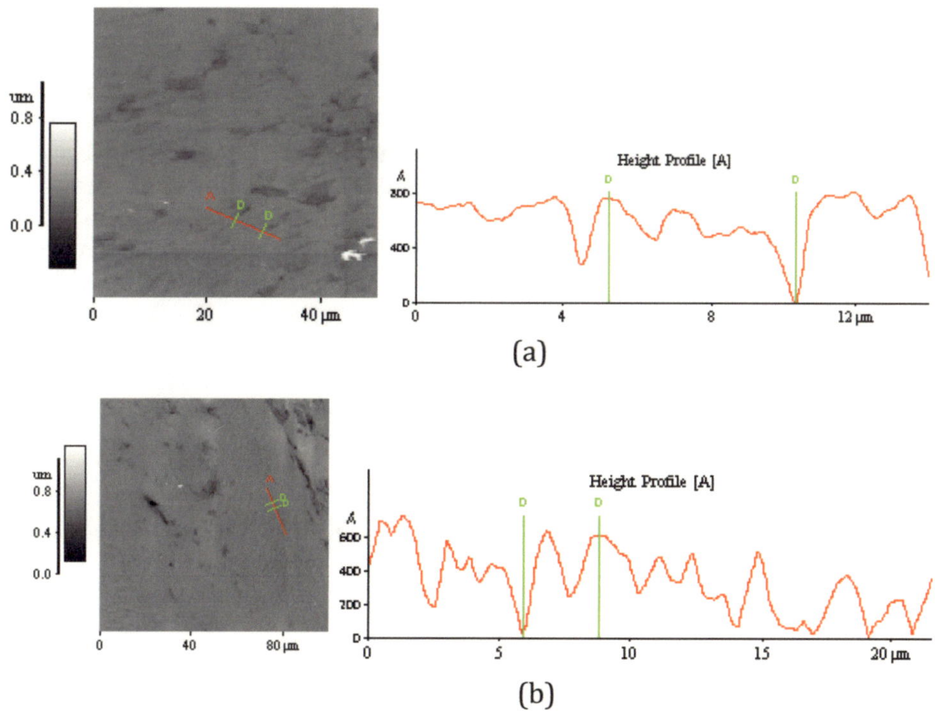

Figure 11.
Atomic force microscopy images and height profiles of microstructural elements shown irradiated, (a) at 80 dpa and (b) at 380 dpa.

Región	Roughness (nm)			Región characteristic:
	Non-irradiated	Irradiated with 80 dpa	Irradiated with 380 dpa	
Dendrític	70.5	215	261	Monophasic
Ínterdendrític	70.5	322	192	Biphasic

Table 2.
Roughness of dendritic and interdendritic regions.

Table 2 shows the results of roughness measurements carried out to evaluate the changes induced by the irradiation dose increase on the lamellae morphology of Ni(Si) and $Ni_3Si_{\text{-monoclinic}}$ phases.

Results analysis shows that the roughness increase rate in the dendritic region is a dose function, an expected result based on the mathematical models of sputtering yield [16]. However, the roughness rate change in the interdendritic region shows a pronounced increase at a dose of 80 dpa, behavior that denotes the effect of Ni(Si) phase preferential erosion, but at a dose of 380 dpa the roughness value decreases markedly. This could be caused by a smoothing mechanism generated by mass transport as a consequence of the irradiation temperature, in order to determine the nature of this surface smoothing mechanism, the geometric factors of the pattern of valleys and ridges generated in the interdendritic region were measured.

Measurements of interdendritic regions' height profiles show a significant change pattern geometry of valleys and ridges formed in that region due to the preferential sputtering generated by the chemical composition of the phases that form the regular eutectic. **Figure 12** shows the diagram of the valley-ridges pattern, as well as the comparative height profile of the dendritic and interdendritic regions at both irradiation doses. **Table 3** shows the average results of the valley- ridges pattern geometric characteristics at both irradiation doses.

When analyzing the results of **Table 3** in conjunction with **Figure 12**, it is concluded that with increasing irradiation dose an additional sputtering mechanism with a geometrical character is present and it is this mechanism that is responsible for irradiated smoothing surface at a dose of 380 dpa. The theoretical foundation of this geometric erosion mechanism was provided by the erosion mechanism proposed by Sigmund [16].

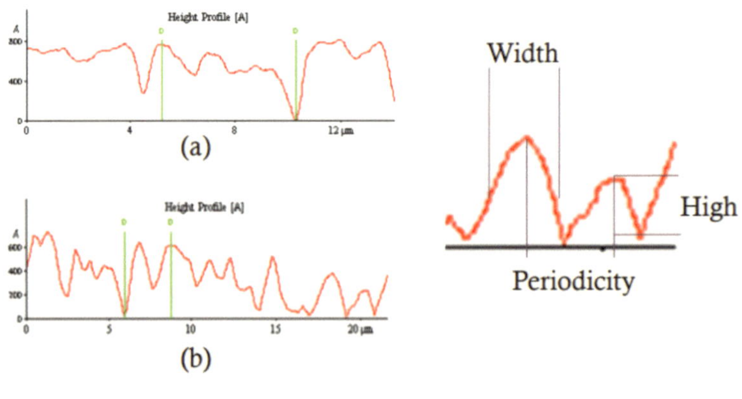

Figure 12.
The valleys and ridges regular pattern diagram, and height profiles at both doses, (a) 80 dpa and (b) 380 dpa.

Dosis (dpa)	80	380
Ridges width (μm)	2.27	1.41
Ridges height (Å)	393	224
Periodicity (μm)	1.2	1

Table 3.
Geometric characteristics of the valleys and ridges regular pattern as a function of irradiation dose.

In conclusion, the interfaces considerably increase the sputtering yield until the ion beam incidence angle reaches a certain critical value at which, the geometric sputtering mechanism is greater than any other sputtering mechanism, and this geometric sputtering promoted the irradiated surface smoothing by a non-diffusional mechanism.

The increased of interdendritic region sputtering yield generated by increased sputtering yield due to grain boundaries sputtering plus interdendritic region sputtering by geometric mechanism promoted the formation of an unevenness between the dendritic and interdendritic region.

4. Surface sputtering applications

The material sputtering as a result of being irradiated with some irradiation type has a large number of practical applications, for example, in thin film deposition and as a material characterization tool. A series of images of the fabrication process of a sample in cross-section for transmission electron microscopy are shown in **Figure 13**. The sample was obtained using a (Focused Ion Beam system) dual beam scanning microscope, this type of microscope has two optical systems, one of electrons and the second of ions, with both systems, it is possible to obtain images with different perspectives. The cutting process is performed by sputtering the selected region using a gallium ion beam.

Figures 13 and **14** show images of the sample selection and cutting processes, both of which are very high-precision processes. The sample sizing process is performed using sputtering of the excess material, depending on the dimensions of the selected

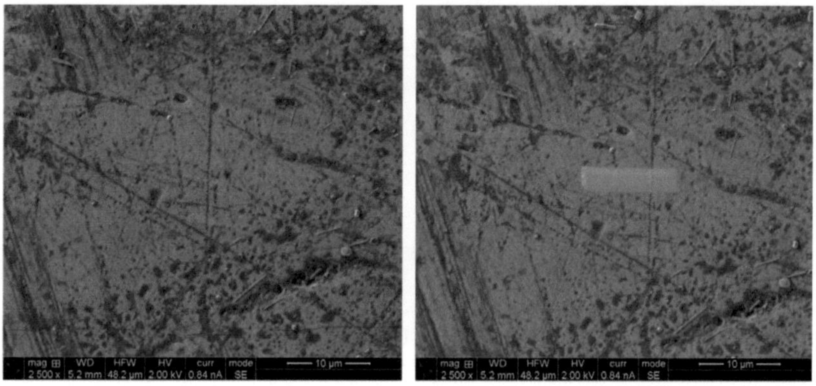

(a) (b)

Figure 13.
The first stage consists of selecting the area from which the sample is required, (a) sampling area and (b) selected sample.

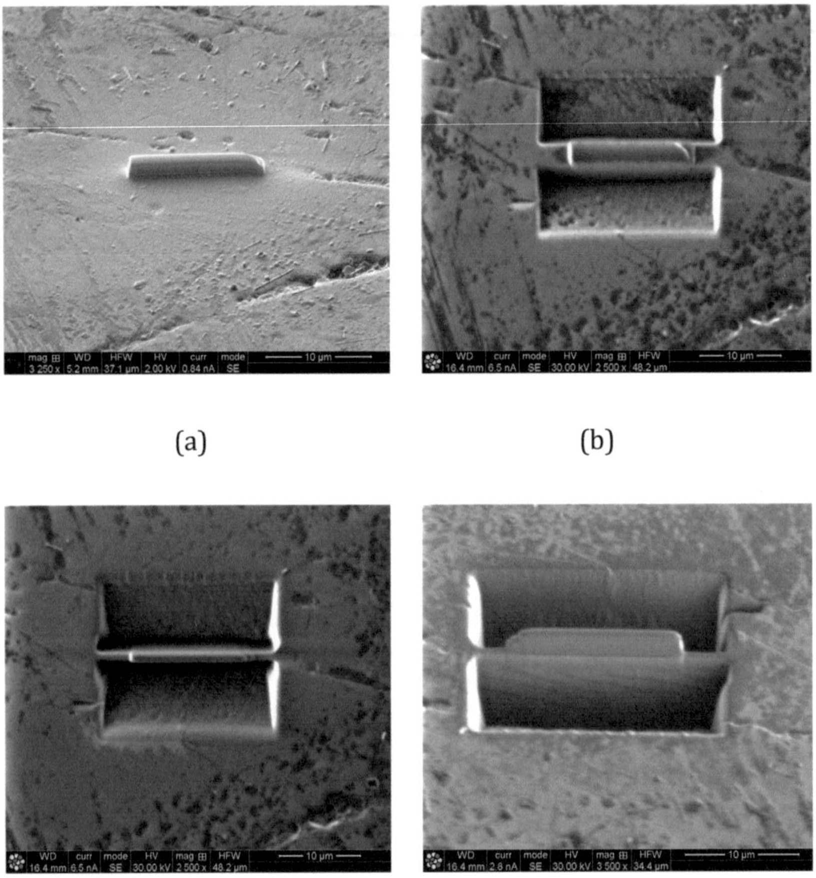

(a) (b)

(c) (d)

Figure 14.
Shows images of the sizing process of the selected sample, (a) sample, (b) cuts to limit the sample, (c) delimitation of the depth and (d) sample ready to be extracted.

area the removal process takes minutes, and the material removed by the sputtering process is evacuated from the chamber via the vacuum system of the microscope. This sizing system can be used on a wide range of materials without causing contamination or modification of the sample microstructure.

Figure 15 shows a series of images of the sample extraction process, as well as the preliminary cuts, welding of the sample with the extraction needle and the extraction process, these steps are very delicate because the sample can be lost in case of hitting the wall of the groove from which it is extracted.

The last stage of the process consists of transferring the sample to a sample holder where the thickness of the sample will be reduced via sputtering to a thickness of 100 nm. **Figure 16** shows the image of the sample placed in the sample holder.

This sampling process allows obtaining samples with very high precision and relatively very fast, which is a consequence of the high spatial resolution of sputtering, which is determined by the precision with which the beam diameter is controlled.

(a) (b)

(C) (d)

Figure 15.
Shows images of the sample extraction process, (a) preparation cuts, (b) approach of the extraction needle, (c) joining of the extraction needle and sample via soldering and (d) sample extraction.

(a) (b)

Figure 16.
Images of the process of mounting the sample in the sample holder to reduce its thickness to about 100 nm.

5. Conclusion

The microstructure changes induced by ion beam irradiation of the irradiated surface depend on the parameters associated with the ion beam, ion type, beam flux, beam energy, and irradiation time, as well as on the surface microstructure, mainly on the phase's chemical composition, the phases distribution on the surface, the surface defects present and the ion beam-surface interaction parameters, mainly the ion beam incidence angle and the mass ratio between target and the ion used to irradiate the material.

As a result of the ion beam—Surface interaction process, several sputtering mechanisms are activated, mainly the chemical one, giving rise to the preferential sputtering of one of the phases present on the surface, the terminal phase Ni(Si) was determined to be preferentially eroded, since it has the highest sputtering yield (see **Table 1**), also when comparing its average molecular weight with the atomic weight of Nickel it is observed that the difference between both is less than 10%, which promotes a greater energy transfer to this phase. On the other hand, it is known that there is a critical local angle of incidence of the ion beam that promotes the activation of the geometric sputtering process, the main characteristic of this geometric mechanism is that it has the potential to modify the topography generated as a result of the preferential erosion of a specific microstructural component, even in an extreme case to produce the smoothing of the surface under irradiation.

Finally, the degree of the changes induced on the irradiated surface are strongly linked to the complexity of the microstructure of the surface of the material used as a target, therefore, the use of sputtering as a tool to develop new materials is conditioned to the microstructural characteristics of the material to be used as target and to the type of ion employed.

Glossary

Atoms implantation: Procedure that consists of incorporating atoms other than that of the base material to modify its characteristic properties.

Crystal lattice: A scheme by which crystal structures are classified according to unit cell geometry. This geometry is specified in terms of the relationships between edge lengths and interaxial angles.

Diffusion: Atoms moving and changing places, i.e., directed or undirected mass transport by atomic motion. Atomic diffusion in solids is far slower than in gases or liquids.

Dislocation: A linear crystalline defect around which there is atomic misalignment. Plastic deformation corresponds to the motion of dislocations in response to an application.

Grain boundaries: The interface separating two adjoining grains having different crystallographic orientations.

Microstructure: The structural features of an alloy (e.g., grain and phase structure) that are subject to observation under a microscope.

Nanostructures: Structure with one or more dimensions at the nanoscale.

Polycrystalline: Refers to crystalline materials that are composed of more than one crystal or grain.

Roughness: is the degree of unevenness of a solid surface below the size scale of its shape or waviness, but above the irregularity of crystal lattice structures. The roughness has an influence on the wettability of a solid.

Segregation: Enrichment of solutes in grains and grain boundaries. An approximation to the segregation tendency of a solute in grain and grain boundary is its apparent solubility: the lower the apparent solubility, the higher the enrichment factor of that element in grain and grain boundary.

Shear stress: The instantaneous applied shear load divided by the original cross-sectional area across which it is applied.

Sputtering: A phenomenon in which microscopic particles of a solid material are ejected from its surface after the material is itself bombarded by energetic particles of a plasma or gas.

Author details

Carlos A. Camacho Olguín[1*], Arturo García Bórquez[2], Carlos A. González Rodríguez[1,3], Héctor Cruz Mejía[1,3] and Marco Solorio Ávila[1]

1 Polytechnic University of the Valley of Mexico, Tultitlan, State of Mexico, Mexico

2 National Polytechnic Institute, Mexico City, Mexico

3 National Autonomous University of Mexico, Mexico City, Mexico

*Address all correspondence to: cachisolguin@hotmail.com

References

[1] Dey S, Dutta AN, Gayathri, Mukherjee PK, Tapatee KR. Characterization of ioninduced microstructural changes in oxygen irradiated Ti–6Al–4V. Radiation Effects and Defects in Solids. 14 Jul 2022;**177**(9-10):972-991. DOI: 10.1080/10420150.2022.2098748

[2] Cui M, Jin P, Shen T, Zhu Y, Pang L, Wang Z, et al. Microstructure change, Nano-Hardness and Surface Modification of CN-G01 Beryllium Induced by Helium Ions. Metals (Basel) [Internet]. 2022;**13**(1):60. Available from: https://www.mdpi.com/2075-4701/13/1/60

[3] Luneville L, Garcia P, Simeone D. Predicting nonequilibrium patterns beyond thermodynamic concepts: Application to radiation-induced microstructures. Physical Review Letters. 2020;**124**(8):085701. DOI: 10.1103/PhysRevLett.124.085701

[4] Camacho-Olguín CA, Garcia-Borquez A, Paz del Angel V, Fuente JAM-DL, González-Rodríguez CA, Ramos-Tercero LM, et al. Formation of the $Ni_{31}Si_{12}$ phase induced by the irradiation of nickel ions on the eutectic α-Ni-Ni_3Si-monoclinic, with a 380 dpa dose. Radiat Eff Defects Solids. 2020;**175**(9-10):925-937. DOI: 10.1080/10420150.2020.1793338

[5] Yano KH, Kohnert AA, Kaspar TC, Taylor SD, Spurgeon SR, Kim H, et al. Dose rate dependent cation & anion radiation enhanced diffusion in hematite. J Mater Chem A Mater Energy Sustain [Internet]. 2022;**10**(45):24167-77. Available from: https://pubs.rsc.org/en/content/articlehtml/2022/ta/d2ta03403d

[6] Belkacemi LT, Meslin E, Crocombette J-P, Radiguet B, Leprêtre F, Décamps B. Striking effect of solute elements (Mn, Ni) on radiation-induced segregation/precipitation in iron-based model alloys. Journal of Nuclear Materials. 2021;**548**:152807

[7] Jiang W, Kovarik L, Zhu Z, Varga T, Bowden ME, Matthews BE, et al. Microstructural evolution and precipitation in γ-LiAlO2 during ion irradiation. Journal of Applied Physics. 2022;**131**(21):215902

[8] Li W, Zhan X, Song X, Si S, Chen R, Liu J, et al. A review of recent applications of ion beam techniques on nanomaterial surface modification: Design of nanostructures and energy harvesting. Small. 2019;**15**(31):e1901820. DOI: 10.1002/smll.201901820

[9] Nastasi M, Mayer JW, Hirvonen JK. Cambridge solid state science series: Ion-solid interactions: Fundamentals and applications: Fundamentals and Applications. Cambridge, England: Cambridge University Press; 1996

[10] Sigmund P. Theory of sputtering. I. sputtering yield of amorphous and polycrystalline targets. Phys Rev [Internet]. 1969;**184**(2):383-416. DOI: 10.1103/physrev.184.383

[11] Camacho-Olguín CA, Rodriguez-Diaz RA, García-Borquez A, Cruz-Mejía H, Camacho-Olguín R. Effect of the energy density deposited by a 3.66-MeV Nickel ion beam on the sputtering yield of a hypereutectic alloy. Radiat Eff Defects Solids. 2022;**177**(9-10):1033-1048. DOI: 10.1080/10420150.2022.2105215

[12] Koslowski HR, Schmitz J, Linsmeier C. Segregation, and preferential sputtering of Cr in WCrY

smart alloy. Nucl Mater Energy. 2020;**22**(100736):100736. Available from: https://www.sciencedirect.com/science/article/pii/S2352179120300120

[13] Barr CM, Chen EY, Nathaniel JE II, Lu P, Adams DP, Dingreville R, et al. Irradiation-induced grain boundary facet motion: In situ observations and atomic-scale mechanisms. Science Advances. 2022;**8**(23). DOI: 10.1126/sciadv.abn0900

[14] SRIM-2013 Manual and Code. Available from: http://www.SRIM.org

[15] Simeone D, Garcia P, Luneville L. Radiation-induced patterning at the nanometric scale: A phase field approach. Materials (Basel). 2022;**15**(9):2991. DOI: 10.3390/ma15092991

[16] Cupak C, Szabo PS, Biber H, Stadlmayr R, Grave C, Fellinger M, et al. Sputter yields of rough surfaces: Importance of the mean surface inclination angle from nano- to microscopic rough regimes. Applied Surface Science. 2021;**570**:151204. DOI: 10.1016/j.apsusc.2021.151204

Chapter 4

Oblique Ar^+ Sputtered SiC Thin Films: Structural, Optical, and Electrical Properties

Divya Gupta, Usha Rani, Rahul Singhal and Sanjeev Aggarwal

Abstract

State-of-the-art experimental facility 200 kV ion accelerator, with energy range of 30–200 kV has been running successfully at Ion Beam Center, KUK. The versatility of this facility lies in providing single charge state and large area irradiation in a single step. In this regard, present work investigates the structural, optical, and electrical behavior of as-deposited and argon-sputtered SiC thin films at varying fluences keeping ion incidence at 50^0. Raman measurements reveal that the opening of sp^2 sites on a-C results in increased sp^3 content in the surface layers. Both FWHM of G peak and I(D)/I(G) ratio decrease with increasing argon fluence. UV-Vis-NIR spectroscopy reveals an increase in the optical absorption and a shifting of absorption edge toward longer wavelengths. I-V characteristics reveal ohmic behavior of all the samples in the voltage range of −5 to +5 V. The conductivity of all the samples is found to decrease with increase in argon ion fluence. The observed structural transformations are attributed to the different degree of sputtering yield of silicon and carbon at different argon ion fluences.

Keywords: silicon carbide, ion irradiation, Raman spectroscopy, UV-Vis spectroscopy, I-V characteristics

1. Introduction

Silicon carbide (SiC), a compound composed of silicon and carbon, has emerged as a remarkable material with a wide range of applications, revolutionizing various industries and research fields. Known for its exceptional mechanical strength and high thermal conductivity to its excellent electrical properties and wide bandgap, SiC has captured the attention of scientists and offers a myriad of possibilities for technological advancements [1–9].

Ion implantation stands as a pioneering technique at the forefront of materials engineering. This technique has opened new avenues for tailoring and enhancing the structural, optical, and electrical characteristics of materials. Interestingly, oblique ion implantation, a specialized technique within ion beam modification, has emerged as a powerful method for tailoring material properties with unprecedented precision. Unlike traditional perpendicular ion implantation, oblique implantation involves the controlled bombardment of a target material at an angle, introducing ions into the surface with an oblique trajectory. This unique approach opens a realm of possibilities, allowing engineers and researchers to engineer intricate structures, manipulate surface morphology, and optimize material characteristics in ways that were once considered challenging or even unattainable [8–17].

Oblique ion implantation offers a distinct set of advantages, enabling the creation of anisotropic material modifications, surface patterning, and enhanced doping profiles. By varying the implantation angle, energy, and ion species, it becomes possible to engineer intricate 2D and 3D surface features effectively on a micro and nanoscale level. These tailored structures find applications across diverse fields, from microelectronics and optoelectronics to biomaterials and sensor technologies. Thus, by carefully controlling the ion implantation parameters, it is possible to engineer SiC with tailored structural and surface properties to meet the requirements of specific applications [9–15].

Thus, the present work investigates the structural, optical, and electrical properties of oblique argon-sputtered silicon carbide surfaces. As a novelty, we study the surface characteristics of unexplored SiC/Si(111) surfaces at oblique argon ion implantation with the aim of quantifying the dependence of structural, optical, and electrical behavior on the ion fluence. The oblique ion beam erosion, causing the removal of the surface target atoms, is the pivotal mechanism in this low-energy region. The investigations of optical and electrical response in combination with the structural modifications taking place in the silicon carbide thin film surfaces have been studied.

2. Experimental details

2.1 Sample preparation

SiC thin films were grown onto Si(111) substrate by radio-frequency (RF) sputtering technique, detailed in our earlier publication [18]. Then, these as-deposited thin film samples were mounted on a substrate holder with the possibility that angle of ion beam incidence could be fixed to 50^0 with respect to the surface normal. For ion beam sputtering experiments, these samples were irradiated with 80 keV Ar^+ ion beam to fluences of 5×10^{17} Ar^+cm^{-2} and 7×10^{17} Ar^+cm^{-2} at incident angle of 50^0 with respect to the surface normal using 200 kV ion accelerator facility available at Ion Beam Center, Kurukshetra University, Kurukshetra. The base pressure in the target chamber was 1.2×10^{-6} Torr. The ion current density during irradiation was kept constant at 2.8 $\mu A\ cm^{-2}$ [19].

2.2 Instrumentation

The structural behavior has been recorded by Raman spectroscopy. Raman spectra of the argon-sputtered films were acquired by micro-Raman spectrometer with a laser excitation wavelength at 532 nm. The optical behavior has been studied by UV–VIS spectrophotometer.

The diffuse reflectance measurements have been performed on Si(111), SiC, and argon-sputtered SiC samples using a Shimadzu UV-VIS-NIR spectrophotometer (UV-3600 Plus), equipped with integrating sphere assembly ISR-603 in the wavelength range of 200–800 nm with a resolution of 0.1 nm. All the reflectance spectra were recorded keeping air as the reference.

Current–voltage (I-V) characteristics and conductivity measurements have been performed by employing Keithley 4200A-SCS parameter analyzer by applying voltage of −5 V to +5 V and measuring the resulting current.

3. Results and discussion

3.1 Structural investigations of RF sputtered SiC over Si(111)

In order to investigate the structural changes after 80 keV argon ion sputtering in SiC/Si(111) surfaces, we have performed Raman scattering in backscattering geometry at room temperature.

Figure 1 displays the room temperature (RT) first-order Raman spectra acquired from a bare Si(111) substrate, the as-deposited SiC film and 80 keV argon ion sputtering at various argon ion fluences, respectively.

Figure 1(a) reveals that Raman spectra of Si(111) consists of one sharp peak at 520 cm^{-1} corresponding to first-order transverse optical phonon (1TO) vibrations of crystalline silicon [1–4]. In addition, the presence of second-order transverse optical mode (2TO) at 929–1027 cm^{-1} has been identified. This validates the crystalline nature of the surface layers in Si(111).

Figure 1(b) exhibits that Raman spectra of SiC RF sputtered over Si(111) display the presence of longitudinal and transverse phonon modes of Si-C bonds in SiC at wavenumber of 972 and 796 cm^{-1}, respectively. Generally, the 3C-SiC single crystal has these two strong phonon modes corresponding to longitudinal optical (LO) and transverse optical (TO) vibrations, whose Raman shifts are 972 and 796 cm^{-1}, respectively [1–4]. The presence of these two phonon modes but with reduced intensity and board linewidth in Raman spectra in **Figure 1(b)** confirms that the SiC layers, analyzed in this work, consist mainly of cubic poly-type structure [1–5]. Due to the

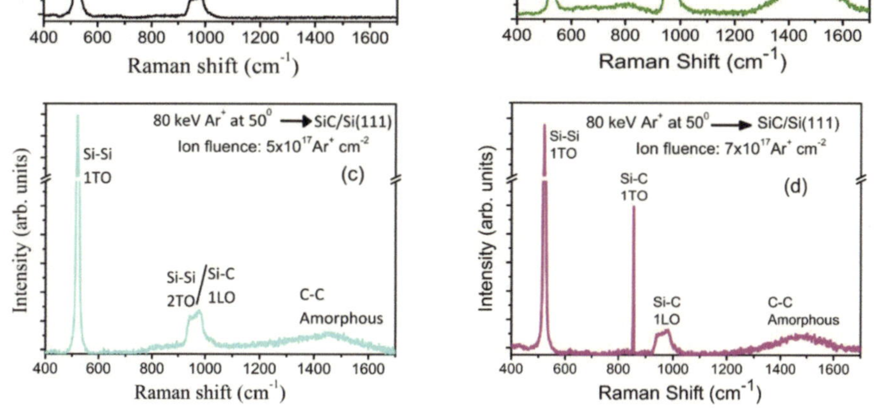

Figure 1.
Raman spectra of (a) bare Si(111) substrate, (b) as grown SiC on Si(111), 80 keV argon-sputtered SiC surfaces for fluence of (c) 5 × 10¹⁷ Ar⁺ cm⁻² and (d) 7 × 10¹⁷ Ar⁺ cm⁻². TA-transverse acoustical, TO-transverse optical, LA-longitudinal acoustical, and LO-longitudinal optical.

broad distribution of these LO and TO phonon modes, we can assume that the lattice of 3C-SiC is amorphous in nature.

Additionally, Raman spectrum (**Figure 1(b)**) shows prominent Si peak at 520 cm^{-1} as observed in Si sample. The peak is detected due to the extended depth of the 532 nm laser light in 3C-SiC. In principle, the penetration depth of a given laser in material depends on the incident wavelength used in Raman measurements. The efficient penetration depth of Ar^+ ion laser having wavelength of 532 nm in 3C-SiC is larger than 83 μm, and its penetration depth in Si is about 0.935 μm [1–8]. Hence, our as-deposited 3C-SiC films of thickness 0.28 μm appear transparent. Thus, the penetrating light is absorbed within about 0.935 μm of the Si substrate, not in the 0.28 μm thick 3C-SiC, whose phonon intensity is much smaller than that of silicon.

Interestingly, selection rule for the (111) surface of a zinc blende crystal allows both LO and TO phonon mode. In our case, the broadband seen in the range 900–1010 cm^{-1}, which we have devoted to LO vibrations of Si-C bonds in SiC is also associated with Si second-order, that is, 2TO Raman scattering [4–12]. As we have already discussed the penetration depth of the 532 nm laser is more than the thicknesses of 3C-SiC films, so, here, LO phonon peak of SiC layer peak is overlapped with the second order Raman spectrum of silicon as evidenced from **Figure 1(b)**. Interestingly, one can observe small humps on both sides of the LO band. These shoulders are attributed to 2TO peak spreading in the wavenumber region of 920–1050 cm^{-1}. This, in turn, complicates the use of LO peak position for drawing the conclusion on SiC structure.

Figure 1(c and **d)** reveals that sputtered SiC surfaces have undergone a substantial change indicative of structural rearrangement. The argon beam interaction has produced severe structural modifications in the surface layers of irradiated specimens. These spectra present three main broad peaks in the 100–550, 550–1100, and 1100–1700 cm^{-1} spectral region corresponding to Si-Si, Si-C, and C-C bonding, respectively.

Similar to as-deposited sample, Raman spectrum in **Figure 1(c** and **d)** also display sharp 1TO peak at 520 cm^{-1} characteristic of predominant crystalline structure of the Si substrate. This is due to the large penetration depth of the 532 nm light in 3C-SiC.

Additionally, Raman spectra display the presence of longitudinal phonon mode of Si-C bonds in SiC at wavenumber of 972 cm^{-1}, which is being overlapped with Si-Si bonding 2TO peak.

To qualitatively asses the structure of 3C-SiC, we have further studied in details the contribution from C-C bonds in as-deposited and 80 keV argon-sputtered SiC thin films. For this, we have deconvoluted the region from 1100 to 1700 cm^{-1} of the Raman spectrum of **Figure 1(b, c** and **d)**. The deconvoluted Raman spectra of SiC thin films within 1100–1700 cm^{-1} are shown in **Figure 2**.

Figure 2 depicts that after Gaussian fitting of the Raman spectra, splitting into two peaks has been seen. One typical peak of amorphous carbon (a-C) centered at 1582 cm^{-1} characteristic of G peak has been observed [13]. The G peak is the result of hybrid C-C bond diffusion of symmetrical stretching vibrating graphite sp^2 carbon [6–8]. Interestingly, one a-C peak at 1463 cm^{-1} also appears. The presence of this peak is attributed to the amorphous carbon content in the sample. Further, this peak presents the incomplete crystallization of carbon clusters [7–19]. Zhang et al. [15] reported the presence of these amorphous carbon bands and ascribed to the mixed sp^2 and sp^3 C-C bonds in amorphous SiC.

Figure 2(b and **c)** depicts that after Gaussian fitting of the Raman spectra, splitting into three peaks has been seen for SiC films following argon ion bombardment at oblique incidence of 50^0 for fluences of 5×10^{17} Ar^+cm^{-2}, 7×10^{17} Ar^+cm^{-2}. Two typical

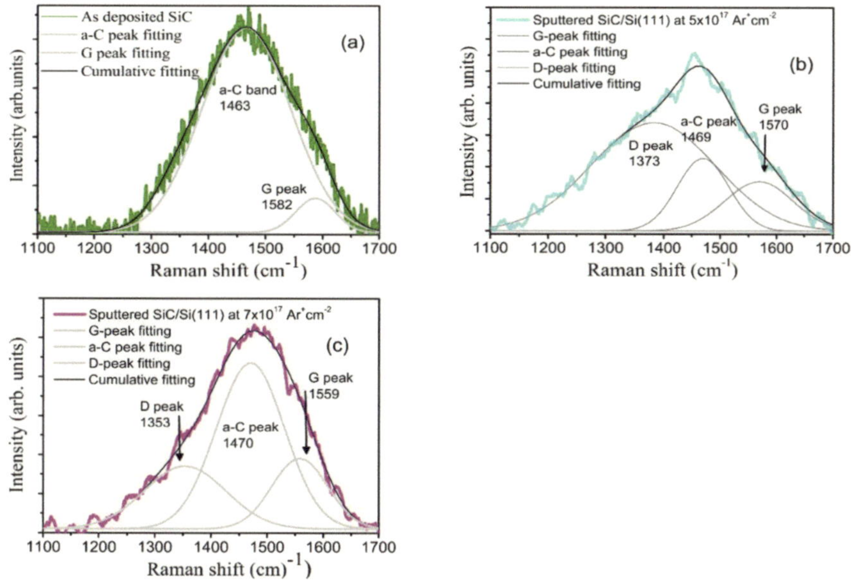

Figure 2.
Raman spectra of (a) as grown SiC on Si(111), 80 keV argon-sputtered SiC surfaces for fluence of (b) 5 × 10^{17} Ar^+ cm^{-2} and (c) 7 × 10^{17} Ar^+ cm^{-2} in the wavenumber region 1100–1700 cm^{-1}.

Specimen SiC/Si(111) Ar^+ cm^{-2}	Center			FWHM			Intensity			I(D)/I(G)
	D Peak	a-C Peak	G Peak	D Peak	a-C Peak	G Peak	D Peak	a-C Peak	G Peak	
0	—	1466	1587	—	155	69	—	2100	352	—
5 × 10^{17}	1373	1469	1570	207	99	121	647	518	326	1.98
7 × 10^{17}	1353	1470	1559	145	122	99	321	851	360	0.89

Table 1.
Raman deconvolution parameters for varying argon ion fluences.

peaks of amorphous carbon (a-C) centered at 1373 and 1570 cm^{-1} characteristic of G peak and D peak have been observed, respectively [19]. Interestingly, one a-C peak at 1469 cm^{-1} also appears. This symmetric peak can be ascribed to the presence of very small and highly disordered carbon clusters. The Raman parameters, that is, peak positions of G and D peak, FWHM of G peak and D peak and their integrated intensities have been determined from deconvolution procedure. These values have been listed in **Table 1**.

From **Table 1**, it can be seen that the position of G peak lies in the spectral region of 1559–1587 cm^{-1}, which is evidence for mixture of sp^2 and sp^3 hybridization. In general, decreasing intensity ratio can be interpreted in terms of corresponding increase in sp^3 content.

Table 1 depicts that G peak lies at 1587 cm^{-1} for as-deposited SiC/Si(111) surfaces. This position of G peak, that is, P(G) moves progressively toward lower wavenumber for SiC surfaces following argon beam exposure at 5 × 10^{17} Ar^+ cm^{-2}. This downshift is attributed to the lowering of C=C vibration modes by the heavier Si atoms in the surface layers. An additional cause of downshifting may be due to higher electronegativity of carbon atoms directly linked to silicon atoms. This direct linkage subsequently reduced the strength of C=C bonds and hence, absorption with lower frequencies. Along with this, linewidth broadening of G peak is observed with respect to the argon fluence. This clearly reveals the formation of more disordered amorphous structure at this stage of sputtering. Interestingly, intensity ratio of D peak to G peak is 1.98 revealing increase in sp^3 content of C-C bonding in the surface layers. Thus, it can be concluded that initial amorphous SiC surfaces are transforming to highly disorder amorphous structure subjected to argon ion irradiation at lowest ion fluence of 5×10^{17} Ar^+cm^{-2}.

With increase in ion fluence from 5×10^{17} to 7×10^{17} Ar^+ cm^{-2}, further shift in G peak position toward lower wavenumber has been seen. In contrast to earlier case, we observed less dispersion in the G peak. Moreover, intensity ratio of D peak to G peak decreases drastically. Further, the fraction of amorphous Si increase at this stage of sputtering. Increasing Si content opens up the sp^2 sites, thereby creating more sp^3 content on the surface.

Hence, Raman measurements reveal that the opening of sp^2 sites on a-C results in increased sp^3 content in the surface layers. Both FWHM of G peak and I(D)/I(G) ratio decrease with increasing argon fluence.

3.2 Optical investigations of RF sputtered SiC over Si(111)

The optical behavior of RF sputtered SiC thin film and argon-irradiated SiC samples have been investigated by means of UV–Vis diffuse reflectance spectroscopy. **Figure 3** presents the diffuse reflectance spectra of as-deposited and argon-sputtered SiC thin films at different fluences.

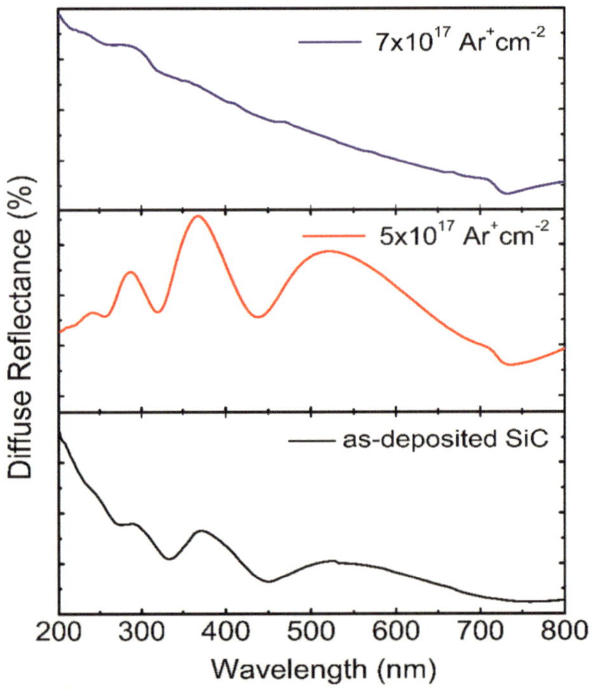

Figure 3.
Diffuse reflectance spectra of as-deposited and argon-sputtered SiC thin films.

It is clear from **Figure 3** that diffuse reflectance decreases uniformly with increase in wavelength and shows prominent peak near 370 nm indicating the presence of fundamental absorption band in this region for as-deposited thin film. Multiple fringes can be clearly seen for SiC thin film sputtered with argon ion fluence of 5×10^{17} Ar^+ cm^{-2}. Surprisingly, for samples sputtered with argon ion fluence of 7×10^{17} Ar^+ cm^{-2}, only one broad hump near 280 nm is observed. These diffuse reflectance spectra are converted into corresponding absorption spectra using the Kubelka-Munk theory.

Figure 4 presents the variation of Kubelka-Munk function as a function of energy for different samples.

Firstly, the square of Kubelka-Munk function was calculated. Thereafter, this quantity is plotted with incident energy. This variation of F(R) with energy as plotted in **Figure 4** is now extrapolated onto the x-axis (energy values). This extrapolation gives the measure of the values of the optical energy gap. The values of optical energy gap for SiC thin film as-deposited and irradiated with varying argon ion fluences are summarized in **Table 2**.

Table 2 reveals that optical energy gap varies nonlinearly with argon ion fluence. An increase in the optical absorption and a shifting of absorption edge toward longer wavelengths has been observed. The observed behavior is attributed to the different degree of sputtering yield at different argon ion fluences and non-stochiometric sputtering of silicon and carbon.

3.3 Electrical investigations of RF sputtered SiC over Si(111)

Figure 5 illustrates the current–voltage (I-V) characteristics of bare Si(111), as-deposited SiC thin films and argon-sputtered SiC thin films.

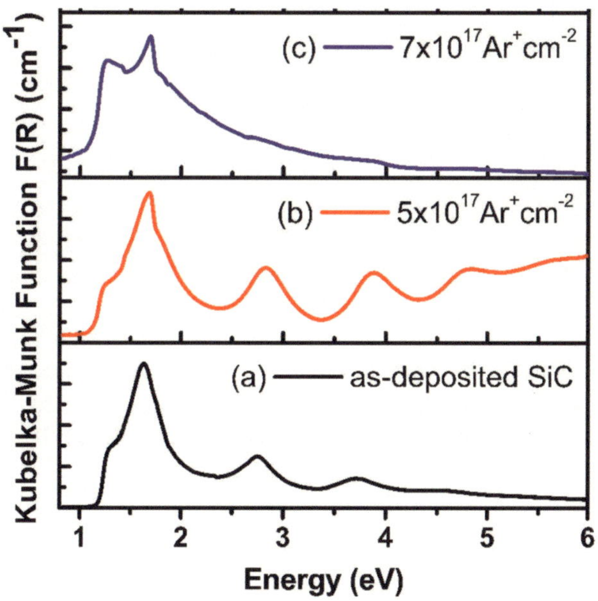

Figure 4.
Kubelka-Munk function F(R) against energy for SiC thin film as-deposited and irradiated with different argon ion fluences.

S. no.	Sample	Optical energy gap (eV)
1.	As-deposited SiC	1.44; 2.49; 3.32
2.	5×10^{17} Ar^{+} cm^{-2}	1.44; 2.58; 3.60
3.	7×10^{17} Ar^{+} cm^{-2}	1.18; 1.37

Table 2.
Values of optical energy gap.

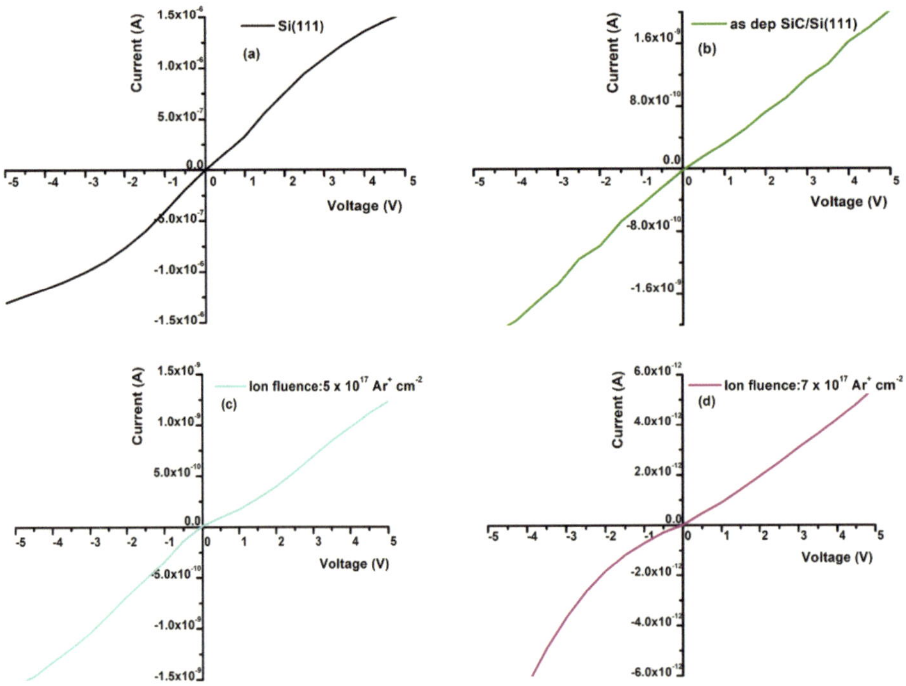

Figure 5.
Current–voltage (I-V) characteristics of (a) bare Si(111) substrate, (b) as grown SiC on Si(111), 80 keV argon-sputtered SiC surface for fluence of (c) 5×10^{17} Ar^{+} cm^{-2} and (d) 7×10^{17} Ar^{+} cm^{-2}.

The current passing through the films exhibited a noticeable reduction transitioning from (a) Si(111) to (b) SiC thin film and then with increase in argon ion fluence (c) and (d). This decrease in current exhibited a linear relationship graphically represented in **Figure 5**. The linear correlation between the applied voltage and the resulting current indicates that the as-deposited and argon-sputtered SiC thin films exhibit ohmic behavior within the designated voltage range of −5 to 5 V.

Moreover, the resistivity of the thin films can be determined by the equation [20–21]:

$$\rho = R\frac{A}{l} \tag{1}$$

In this equation, ρ represents the resistivity, R stands for the resistance, A denotes the surface area of the SiC thin films, and l represents the distance between the two measurement probes.

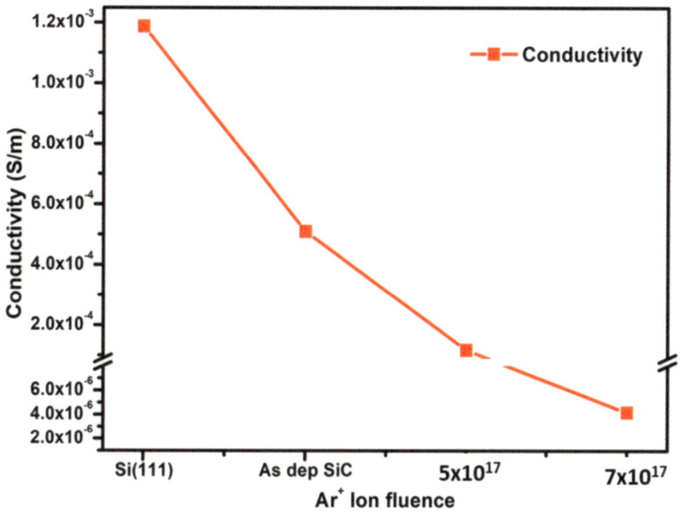

Figure 6.
Variation of conductivity of SiC thin films with Ar^+ ion fluence.

Interestingly, using these values of resistivity, conductivity can be calculated as

$$\sigma = \frac{1}{\rho} \tag{2}$$

Here, σ signifies the conductivity of the thin films and ρ corresponds to the resistivity of the SiC films. **Figure 6** represents the variation of conductivity as a function of argon ion fluence.

Interestingly, the conductivity of thin SiC thin films was found to decrease with increasing argon ion fluence. The increase in resistivity and decrease in conductivity of SiC thin films with increasing ion fluence can be attributed to the introduction of defects and disruptions in the crystal lattice. The ion beam bombardment leads to the creation of vacancies, interstitials, and lattice distortions in the SiC thin films. These defects act as scattering centers for charge carriers (electrons or holes), hindering their mobility and thereby increasing resistivity. The mobility of charge carriers is reduced due to an increase in scattering leads to a decrease in the conductivity of SiC thin films. In the case of the Si(111) wafer, the observed highest conductivity could be due to its single-crystalline structure and minimal defect density compared to the irradiated SiC thin films [22].

The increase in the optical band gap value with ion fluence suggests that the material's electronic structure is being modified. The ion bombardment can induce changes in the electronic states of the material, including the creation of localized energy levels within the band gap. These levels can trap charge carriers, reducing their mobility and contributing to a decrease in conductivity.

4. Conclusions

In summary, the present work investigates the structural, optical, and electrical behavior of as-deposited and argon-sputtered SiC thin films at varying fluences keeping ion incidence at 50^0. Raman studies reveal increased sp^3 content in the surface layers. Both FWHM of G peak and I(D)/I(G) ratio decrease monotonically with increasing argon fluence. Further, an increase in the optical absorption and a shifting of absorption edge toward longer wavelengths has been observed. The I-V Characteristics reveal that the as-deposited and argon-sputtered SiC thin films exhibit ohmic behavior within the designated voltage range of −5 to 5 V. The observed structural behavior is attributed to the different degree of sputtering yield of silicon and carbon at different argon ion fluences. Hence, we propose that these structurally altered SiC thin films find more practical applications in advanced optoelectronics, photovoltaics, media recording, and storage devices.

Acknowledgements

This research is primarily supported by the Department of Science and Technology (DST), New Delhi by funding major research project for utilizing 200 kV ion accelerator and related characterization facilities. The authors are highly thankful to the Ministry of Human Resource and Development (MHRD) for RUSA 2.0 grants for the Center for Advanced Material Research (CAMR) at Kurukshetra University. The authors are grateful to Dr. Dinakar Kanjilal, Raja Ramana Fellow, and Dr. Sundeep Chopra, Inter University Accelerator Center (IUAC), New Delhi for fruitful discussions.

Author details

Divya Gupta[1*], Usha Rani[1], Rahul Singhal[2] and Sanjeev Aggarwal[1]

1 Department of Physics, Ion Beam Centre, Kurukshetra University, Kurukshetra, India

2 Department of Physics, Malviya National Institute of Technology (MNIT), Jaipur, India

*Address all correspondence to: divyagupta2017@kuk.ac.in

References

[1] Zhang L, Jiang W, Pan C, et al. Raman study of amorphization in nanocrystalline 3C–SiC irradiated with C+ and He+ ions. Journal of Raman Specroscopy. 2019;**50**(8):1-8

[2] Wasyluk J, Perova TS, Kukushkin SA, Osipov AV, Feoktistov NA, Grudinkin SA. Raman investigation of different polytypes in SiC thin films grown by solid-gas phase epitaxy on Si (111) and 6H-SiC substrates. Materials Science Forum. 2010;**645-648**:359-362

[3] Nakashima S, Harima H. Raman investigation of SiC polytypes. Physica Status Solidi A: Applications and Material Science. 1997;**162**:39

[4] Xu Z, He Z, Song Y, Fu X, Rommel M, Luo X, et al. Topic review: Application of Raman spectroscopy characterization in micro/nano-machining. Micromachines. 2018;**9**:361

[5] Gupta D, Chawla M, Singhal R, Aggarwal S. Nanoscale structural defects in oblique Ar^+ sputtered Si(111) surfaces. Scientific Reports. 2019;**9**:15531

[6] Feng ZC, Zhao JH. Silicon Carbide: Materials, Processing and Devices. New York: Taylor & Francis Books; 2004. p. 256

[7] Gupta D, Kumari R, Umapathy GR, Singhal R, Ojha S, Sahoo PK, et al. Self-assembled nano-dots structures on Si (111) surfaces by oblique angle sputter-deposition. Nanotechnology. 2019;**30**:385301-385312. DOI: 10.1088/1361-6528/ab273a

[8] Chen et al. Adducing crystalline features from Raman scattering studies of cubic SiC using different excitation wavelenghs. Journal of Physics D: Applied Physics. 2016;**50**:115102

[9] Cheng Y, Huang X, Du Z, Xiao J. Effect of sputtering power on the structure and optical band gap of SiC thin films. Optical Materials. 2017;**73**:723-728

[10] Paik N. Raman and XPS studies of DLC films prepared by a magnetron sputter type negative ion source. Surface and Coating Technology. 2005;**200**:2170-2174

[11] Xiao JR, Xu H, Guo AM, et al. Study on FN-DLC thin films: (II) Effect of radio frequency power on the optical band gap of the thin films. Acta Physica Sinica. 2007;**56**:1809-1814

[12] Daouahi M, Rekik N. Effect of substrate temperature on (micro/nano) structure of a-SiC: H thin films deposited by radio-frequency magnetron sputtering. Journal of Physical Chemistry C. 2012;**116**:21018-21026

[13] Daouahi M, Omri M, Kerm AGY, et al. An investigation on the effect of high partial pressure of hydrogen on the nanocrystalline structure of silicon carbide thin films prepared by radio-frequency magnetron sputtering. Spectrochimica Acta, Part A: Molecular and Biomolecular Spectroscopy. 2015;**136**:1409-1417

[14] Racine B, Ferrari AC, Morrison NA, et al. Properties of amorphous carbon silicon alloys deposited by a high plasma density source. Journal of Applied Physics. 2001;**90**:5002-5012

[15] Ferrari AC, Robertson J. Interpretation of Raman spectra of disordered and amorphous carbon. Physical Review B. 2000;**61**:14095-14107

[16] Fraga MA, Massi M, Oliveira IC, Maciel HS, dos Santos Filho SG,

Mansano RD. Nitrogen doping of SiC thin films deposited by RF magnetron sputtering. Journal of Materials Science: Materials in Electronics. 2008;**19**:835-840

[17] Chehaidar A, Carles R, Zwick A, Meunier C, Cros B, Durand J. Chemical bonding analysis of a-SiC : H films by Raman spectroscopy. Journal of Non-Crystalline Solids. 1994;**169**:37

[18] Gupta D, Umapathy GR, Singhal R, Ojha S, Aggarwal S. Ripple patterns over oblique Ar+ sputtered SiC/Si(1 1 1) surfaces: Role of preferential sputtering. Material Letters. 2022;**308**:131011

[19] Gupta D, Aggarwal S, Sharma A, Kumar S, Chopra S. 200 kV ion accelerator facility at Kurukshetra University, India. Material Letters. 2022;**308**:131294

[20] Javed A et al. Thickness dependent structural, electrical and optical properties of cubic SnS thin films. Materials Chemistry and Physics. 2020;**246**:122831

[21] Hassanien AS, Akl AA. Effect of Se addition on optical and electrical properties of chalcogenide CdSSe thin films. Superlattices and Microstructures. 2016;**89**:153-169

[22] Gupta D, Kumar R. Ion implantation as an approach for tuning of electronic structure, optical, morphological and electrical transport properties of sputtered molybdenum disulfide thin films. Materials Science in Semiconductor Processing. 2023;**158**:107326

Chapter 5

Electron Beam Processing of Biological Objects and Materials

Ulyana Bliznyuk, Aleksandr Chernyaev, Victoria Ipatova, Aleksandr Nikitchenko, Felix Studenikin and Sergei Zolotov

Abstract

The research explores a wide range of applications for electron accelerators in industrial irradiation processing. It also compares the physical properties of electron beams, dose ranges, and methods used for irradiation of polymers, medical items, transplantology objects, pharmaceuticals, and foods. Moreover, the study discusses the depth dose non-uniformity in objects irradiated with accelerated electrons. The research also highlights the dependency of geometry, density, and chemical composition of the object on the dose distribution. Another focus of the study is computer simulation of electron irradiation method, encompassing all physical and technical parameters to assess the dose distribution throughout the irradiated objects, since without knowing the precise electron beam spectrum, it is impossible to accurately reconstruct the dose distribution throughout the objects. Considering that the beam spectrum cannot always be identified, especially for industrial accelerators, the study presents algorithm for reconstructing the dose distribution in irradiated objects. The final part of the research provides methods for increasing the dose uniformity throughout objects irradiated with electron beams.

Keywords: beam spectrum reconstruction, dose distribution, electron beam spectrum, increasing dose uniformity, irradiation processing, modifier plates

1. Introduction

Irradiation with accelerated electron beams is a convenient all-purpose technology for the processing of various biological objects and materials since it offers undeniable advantages, such as the ability to vary the beam intensity and penetration depth of electrons, minimal changes in the temperature and pressure, as well as the absence of negative effects of chemical compounds, which allows to solve a wide range of tasks, from plant growth stimulation to increasing the wear resistance of metals [1]. Electron beam irradiation of objects is enabled by the transfer of the energy to molecules and atoms of substance of irradiated object. As electrons act on the substance, ionization, and excitation of atoms and molecules lead to physical and chemical reactions that change the properties of the object. One of the main factors that determine the effect of accelerated electrons on the substance is the irradiation dose, which is the ratio between the energy absorbed by the volume and its mass [2, 3]. The research discusses the application of electron beams for processing biological and non-biological objects and establishes the dose ranges and physical properties of electron beams to solve various tasks.

Since objects and materials have diverse properties, the dose distribution throughout irradiated objects differs considerably, even if the same irradiation method is applied, depending on the chemical composition, density, and shape of the objects. The research discusses the influence of physical and chemical properties as well as electron energy on the distribution of absorbed dose over the volume of object.

The dose distribution in various objects can be simulated using transport codes [4], provided that all physical and technical parameters of the irradiation method are accurately reproduced, taking into account the individual properties of biological objects and materials. The study compares different transport codes used for irradiation simulation and presents an algorithm for simulating irradiation exposure using the GEANT4 tool kit [5–8], which is by far the most accurate tool for obtaining the absorbed dose distribution in objects of any geometry and chemical composition when irradiated with accelerated electrons.

Considering that it is not always possible to obtain the beam energy spectrum from manufacturers of electron accelerators, it is crucial to have algorithms for reconstructing the beam spectrum of accelerators used by irradiation facilities [9–14]. The study presents an algorithm to reconstruct the electron beam spectrum and the depth dose distribution in an object of any chemical composition and shape, based on the experimentally measured absorbed dose distribution in any known substance.

Since biological objects have complex biochemical compositions and geometry, they are highly sensitive to the lack of dose uniformity [15–17]. The research presents a method of electron beam modification using aluminum modifier plates to significantly improve the irradiation dose uniformity, which makes it possible to increase the spectrum of biological objects irradiated with electron beams.

2. Electron beam processing in industrial irradiation

Electron beam irradiation is an environmentally friendly technology that uses electron energy to initiate a range of physical, chemical, and biological effects without the use of chemical compounds, high temperature, and pressure. The high kinetic energy and penetrating power of electrons offer significant advantages over the traditional methods used for processing biological objects and materials [1, 18, 19].

An important advantage of accelerated electrons over other radiation sources is their ability to vary the beam current and electron energy [20]. By varying the beam current, it is possible to change the intensity of radiation and, consequently, the dose rate absorbed by the treated object. Varying the energy of electrons allows to control the depth of electron penetration throughout the object, depending on the purpose of irradiation treatment.

Electron accelerators are classified according to the energy levels of electrons they generate [1, 21]. Low-energy electron accelerators with energies ranging from 0.08 to 0.8 MeV are commonly used for the treatment of surfaces with a mass thickness of up to 0.2 g/cm^2. Medium-energy electron accelerators, which generate 0.8–5 MeV electrons, are used for the treatment of objects at the depth not exceeding 0.5 g/cm^2. High-energy electron accelerators with energies of 5–10 MeV are applied for sterilization of medical supplies and instruments, biological materials used in transplantation, food processing, treatment of biowaste, decomposition of industrial effluents, as well as sewage and greywater treatment since the electron beams penetrate the object at the depth of up to 5 g/cm^2.

Flexibility of electron accelerators allows using electron beams in diverse areas, such as crosslinking, curing, and grafting of polymers and composites; modification of the material surface; water purification; improvement of material properties; sterilization of medical supplies and instruments; and irradiation of food products [3].

Table 1 shows the applications of electron beam processing, the combinations of the energy ranges and dose ranges that are used in each application, as well as the beneficial effects used in industry.

Application	Irradiated objects	Energy (MeV)	Doses (kGy)	Purpose
Modification of polymers [21, 28–32]	Polyolefins: polyethylene, polyvinylchloride, ethylene-propylene rubber, polyvinylidene fluoride, and ethylene tetrafluoroethylene copolymer	0.3–5	50–400	Wire and cable insulation
		0.5–3	50–150	Heat-shrinkable materials with a "memory" effect
	Elastomers (automobile tires); Rubbers (rubber gloves, roofing, and waterproofing materials)	0.8–1	1–200	Crosslinking
	Materials containing oligomers and monomers (adhesives, cellulose, varnishes, inks, films, concrete)	0.15–0.4	15–50	Polymerization and Grafting
	Composites (automobile and aircraft components)	1.2–1.5	150–250	
	Hydrogels: polyvinyl alcohol, polyacrylamide, polyvinylpyrrolidone, polyethylene oxide, and methylcellulose	3–10	25–50	
	Materials containing polytetrafluoroethylene; cellulosic waste (sawdust, shavings, straw, small chips, etc.); Teflon waste; Rubber, Cellulose, and Polypropylene materials	0.1–12	500–1500	Decomposition
Sterilized medical supplies [29, 33, 34]	Disposable medical equipment (syringes, needles, masks, surgical gloves, etc.); Packaging material (droppers, Petri dishes, pipette bottles, blood collection tubes); Pharmaceuticals (eye ointments and drops, ointments for burns, saline solutions)	5–10	10–35	Disinfection/ sterilization
Preserving foods [15, 35–38]	Potatoes, onions, garlic, carrots, etc.	up to 10 MeV	0.05–0.2	Inhibiting sprouting
	Seeds, spices, fruits		0.15–0.5	Sterilization for pest management
	Fruits, vegetables		0.5–1	Suppressing ripening
	Meat, fish, poultry, seafood, canned food		1–4	Extending shelf-life by suppressing mold, bacteria, and fungi growth
			3–7	Killing pathogenic bacteria

Application	Irradiated objects	Energy (MeV)	Doses (kGy)	Purpose
	Food additives, hospital diets, emergency rations, space food		10–70	Sterilization
Pollution control [39, 40]	Natural sources of water (rivers, lakes, reservoirs, artesian wells)	0.3–2	0.2–1	Cleaning of industrial wastewater, flue gases, and solid waste
	Wastewater and sewage sludge		0.4–4	
	Gases containing SO_2 and NO_x, where x = 1, 2.		10–20	
	Hospital and airport waste, activated carbon regeneration, and contaminated soil cleanup		1000–1500	
Metals and mineral processing [41, 42]	Gems and minerals	2–50	25–1500	Crystalline and glass coloring/discoloring
	Semiconductors (diodes and thyristors)	9–13	0.1–100	Improvement of properties

Table 1.
Electron beam application and operating parameters [22–27].

Irradiation with continuous and pulsed electron beams has become the main tool for obtaining various polymers in free-radical and ionic polymerization reactions. Since crosslinking enables the formation of chemical bonds between molecular chains and the creation of three-dimensional structures, it is commonly used to improve the physical characteristics of the polymer, including heat resistance [21].

Crosslinking at industrial accelerator with the power up to 350 kW is performed with electron energy ranging from 0.5 MeV to 3 MeV, and beam current from 50 mA to 100 mA. Modification of polymers uses the dose ranging from 1 kGy to 400 kGy depending on the desired effect [25]. Polymerization reaction of low molecular weight compounds with free radicals occurring in the process of curing due to indirect action of accelerated electrons in polymerizing monomers is enabled by electrons with the energy of up to 300 keV [31]. Graft polymerization, which involves grafting various monomers onto a polymer chain to give the polymer properties of the monomer, is found to be efficient at 0.3–0.5 MeV [32]. Irradiation with electron beams with energies of up to 0.3 MeV is used for the synthesis of nanocomposites for structural and magnetic applications, nanogels and hydrogels for drug delivery systems, and for the synthesis of membranes for medical and industrial applications [30, 31].

Electron-beam melting with 0.01–0.3 MeV electrons is utilized to produce complex, intricate geometries with excellent mechanical properties by selectively melting and fusing metal power particles [26]. This technique is used in airspace, medical, and automatic industries for manufacturing high-precision components since it allows to change the composition, morphology, and hardness of metals as well as improve wear resistance [27].

Irradiation to suppress microbial contamination in biological and non-biological objects requires a more complex approach since the desired effect is achieved with a precise combination of the absorbed dose, dose rate, and electron energy [37, 38]. Moreover, the dose range for biological objects is commonly narrower compared with metals, minerals, and other non-organic objects due to their chemical complexity.

Sterilization of food products, medical items, and materials used in transplantology is carried out using electron beams with energies from 3 to 10 MeV and a power of up to 50 kW [34, 35]. In food irradiation, varying beam penetration depth allows to solve a range of tasks, from sprout inhibition with the doses of up to 0.2 kGy to food sterilization with the doses of up 50 kGy [32–35]. The required absorbed dose increases to 35 kGy in the treatment of medical devices for inactivation of viruses and microorganisms [29, 33].

Treatment of drinking water and wastewater is carried out using accelerated electrons with the energy of 0.3–1 MeV [39, 40]. While for drinking water treatment doses up to 1 kGy are applied, biowaste treatment for inactivation of a wide range of viruses and bacteria is carried out at doses up to 1 MGy, which is the maximum dose used to control contamination of organic objects [24].

3. Depth dose distribution in biological objects irradiated with accelerated electrons

Irradiation of biological objects is particularly sensitive to depth dose distribution due to the nature of such objects, different physical and chemical composition as well as a lack of homogeneity and complex geometry. All these factors have an impact on the irradiation uniformity and the efficiency of irradiation processing.

In treatment with accelerated electrons, non-uniformity of irradiation is inevitable due to the nature of the depth dose distribution throughout the irradiated object. The ratio of the minimum value of absorbed dose D_{min} to the maximum value of absorbed dose D_{max} in the object volume is commonly used as a criterion of irradiation dose uniformity [1]. Different categories of biological objects, such as transplantology objects, pharmaceuticals, and food products, require uniformity of at least 80% [1, 38]. Considering that achieving an irradiation uniformity of more than 80% for the objects with a mass thickness of more than 2 g/cm^2 is a challenging task, it is necessary to take into account such factors as packaging filling irregularity, as well as geometry, structure, chemical composition, and density of irradiated objects [15, 43].

This part will discuss the factors, which affect the depth dose distribution in the biological object irradiated with accelerated electrons.

3.1 Electron beam energy

One of the key factors that lead to the non-uniformity of electron beam irradiation is the character of depth dose distribution, which depends on the energy of electrons. **Figure 1a** and **b** shows the dependency of the relative absorbed dose on the penetration depth in water parallelepiped irradiated with 4 MeV, 6 MeV, 8 MeV, and 10 MeV, calculated using the GEANT4 toolkit [44]. **Figure 1a** shows that the higher the electron energy, the deeper the penetration of electrons into the object is and the lower maximum dose is achieved during irradiation.

To illustrate the dependency between the electron energy and depth dose distribution, let us introduce the following parameters of the absorbed depth dose distribution:

L_{max} is the distance from the surface of the object with the maximum absorbed dose;

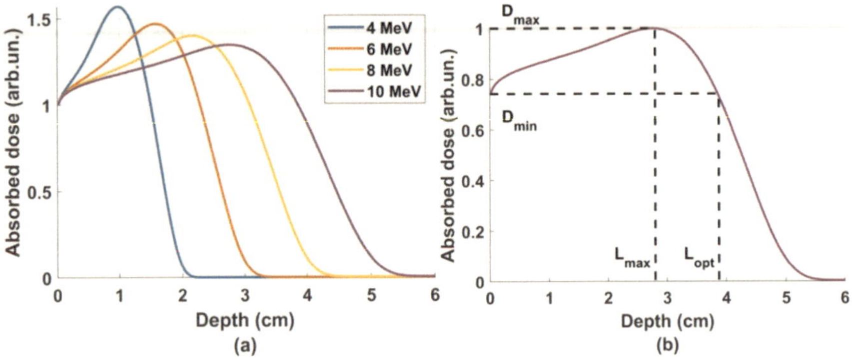

Figure 1.
(a) Depth dose distribution in water parallelepiped irradiated with 4 MeV, 6 MeV, 8 MeV, and 10 MeV electrons; (b) Parameters of absorbed dose distribution in water irradiated with 10 MeV electrons.

L_{opt} is the optimal distance from the surface of the object at which the dose value is equal to the surface dose;

$K = \frac{D_{min}}{D_{max}}$ is the dose uniformity coefficient, which is the ratio of the minimum value of absorbed dose D_{min} to the maximum absorbed dose D_{max} in the object.

As can be seen from **Figure 1b** when the water parallelepiped is irradiated with 10 MeV electrons, L_{max} is 28.0 mm and L_{opt} is 38.75 mm, and the irradiation uniformity coefficient is 0.72. With an increase in electron energy from 4 MeV to 10 MeV, L_{max} increases from 10.25 mm to 27.5 mm, and L_{opt} increases from 15 mm to 38.75 mm (**Figure 1a**). Therefore, varying the beam energy allows to change the dose uniformity of the irradiated object.

3.2 Density of the irradiated object

Considering that biological objects vary in terms of density and density distribution, it is important to investigate the impact of the object density on the parameters of the absorbed depth dose distribution, such as the optimal distance L_{opt} and dose uniformity K. While the irradiation of biological objects is performed with 4–10 MeV electrons, it is feasible to estimate the depth dose distribution coefficients for the objects irradiated within this energy range having the density from 0.3 g/cm^3 to 1.6 g/cm^3, which is similar to that of biological objects at industrial irradiation facilities. **Figure 2a** shows that dose uniformity K varies from 0.62 to 0.72 and practically does not depend on the density of the irradiated object for the water parallelepipeds with densities ranging from 0.3 g/cm^3 to 1.6 g/cm^3.

Figure 2b shows the dependency of optimal distance L_{opt} on the electron energy for irradiation of parallelepipeds with densities of 0.3 g/cm^3, 0.6 g/cm^3, 1.0 g/cm^3, and 1.6 g/cm^3. As it can be seen, the higher the energy of electrons, the greater the value of L_{opt}, which means that at higher energies a greater dose uniformity can be achieved for the objects of greater thickness. At the same time, the lower the density of the irradiated object, the greater the growth rate of L_{opt} value with an increase in electron energy.

According to the computer simulation using the GEANT4 toolkit and MATLAB, the analytical interdependencies of electron energy, dose uniformity K, optimal distance L_{opt}, and the object density can be expressed as follows [45]:

$$L_{opt}[\text{cm}] = 4\left[\frac{\text{cm}^4}{\text{MeV} * \text{g}}\right] \times \rho^{-0.96} \times E[\text{MeV}] - 1.59\left[\frac{\text{cm}^4}{\text{g}}\right] \times \rho^{-0.46}, \quad (1)$$

$$K = 0.01\left[\text{MeV}^{-1}\right] \times E\left[\text{MeV}\right] + 0.57, \quad (2)$$

where ρ is the object density $\left[\frac{\text{g}}{\text{cm}^3}\right]$, E is electron energy. These dependencies were obtained with a maximum interpolation error of no more than 2%.

3.3 The material of irradiated object

The material of irradiated object has an impact on the distribution of absorbed doses throughout the object. This can be explained by the difference in the number of electrons in electron shells of various atoms, which determines the nature of interaction between the accelerated electrons and matter. **Figure 3** shows the dependencies of the relative absorbed dose distribution on the penetration depth of 10 MeV electrons in aluminum, graphite, water, and polypropylene, which have different effective charge Z_{eff}.

Table 2 presents the optimal distances L_{opt} for the object made of aluminum, graphite, water, and polypropylene irradiated with 10 MeV electrons, as well as the irradiation uniformity throughout the object made of different materials, which is determined for the optimal distance L_{opt}.

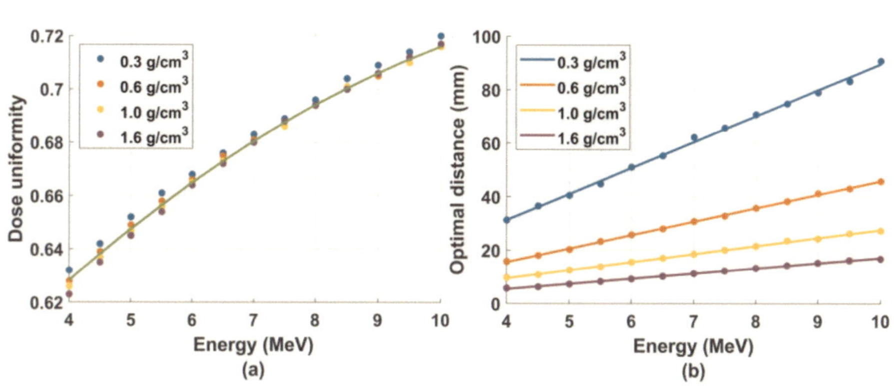

Figure 2.
(a) Dependency of the dose uniformity K in water parallelepiped with the density ranging from 0.3 g/cm³ to 1.6 g/cm³ on the electron energy and the function approximating the calculated dependency (green line); (b) Dependency of L_{opt} on the electron energy for irradiated water parallelepiped with the different density and the functions approximating the calculated dependencies (solid lines).

Material	ρ (g/cm^3)	Z_{eff}	$\rho \cdot Z_{eff}$ (g/cm^3)	K (rel.un.)	L_{opt} (mm)
Aluminum	2.7	13	35.1	0.65	14
Graphite	1.7	6	10.2	0.72	19
Water	1.0	7.4	7.4	0.72	37
Polypropylene	0.9	5.4	4.9	0.75	42

Table 2.
The parameters of absorbed dose distribution in different materials.

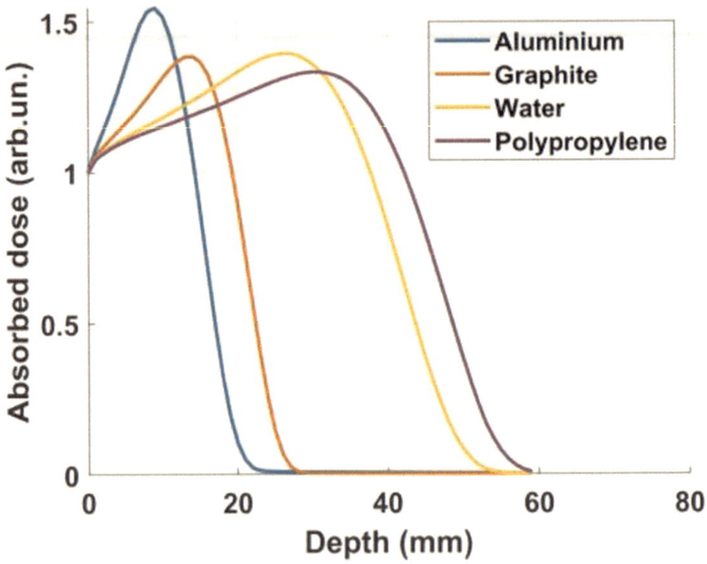

Figure 3.
Depth dose distribution in aluminum (Al, Z_{eff} = 13), graphite (C, Z_{eff} = 6), water (H_2O, Z_{eff} = 7.4), and polypropylene (C_2H_4, Z_{eff} = 5.4) irradiated with 10 MeV electrons.

Electrons with the energies up to 10 MeV, which is the maximum energy used in radiation treatment of biological objects, lose their energy due to ionization losses. It can be seen from **Figure 3**, the higher its effective charge Z_{eff} and density, the more energy is lost by an electron penetrating an equal depth in different materials, according to the Bethe-Bloch formula [46]. Moreover, with an increase in the effective charge Z_{eff} the maximum penetration depth of electrons throughout the object decreases, at the same time causing the surface absorbed dose to decrease compared to the maximum value. Therefore, the material with a higher electron density tends to have a lower irradiation uniformity compared with the material with a lower electron density, as **Table 2** shows.

3.4 The geometry of irradiated object

Biological objects exposed to accelerated electrons, such as foodstuffs and materials used in transplantology may have complex geometry that can be dramatically different from parallelepiped. A parallelepiped is regarded as a perfect shape for irradiation because its simple geometry allows electrons to penetrate the object perpendicularly to its surface. Exposure of more complex geometries, such as a sphere, an ellipsoid, or a cylinder, to electron beams, however, does not permit the perpendicular penetration of electrons into the surface layer, making the depth dose distribution irregular and less predictable. **Figure 4** is a 3D-color representation of the absorbed dose distribution over a parallelepiped with the edge of 6 cm, a ∅ 6 cm sphere, and a ∅ 6 cm cylinder simulated as water phantoms as they are unilaterally irradiated with 10 MeV electrons.

As it can be seen from **Figure 4**, the distance at which the absorbed dose reaches its maximum changes with the change in the object shape. In the case of the parallelepiped, as the electrons start penetrating the volume, the dose increases slightly but when the depth exceeds 2 cm the dose drops to zero at the distance of 5 cm from the surface

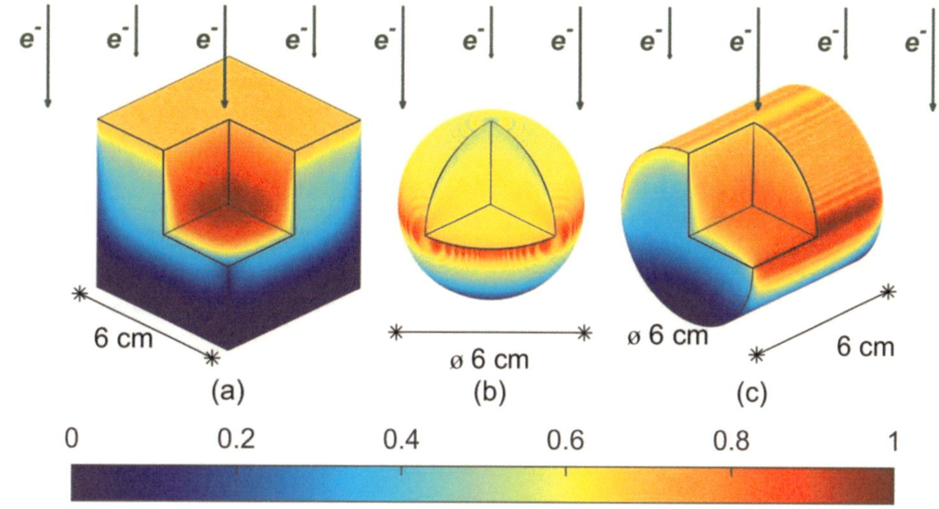

Figure 4.
3D-color map of absorbed dose distribution throughout the water phantoms: (a) parallelepiped with the edge of 6 cm, (b) 6 cm sphere, and (c) 6 cm cylinder with the height of 6 cm during unilateral 10 MeV electron irradiation.

(**Figure 4a**). It is interesting that the layers of the parallelepiped close to the edges are underexposed. When a sphere is irradiated with 10 MeV electrons the layers of the sphere close at the equator are clearly overexposed (**Figure 4b**). In the cylinder, overexposure can be observed on the lateral sides of the surface while the character of depth dose distribution is similar to that of the parallelepiped (**Figure 4c**). When irradiating from one side, all the shapes are partly underexposed the maximum electron penetration depth is lower compared to the linear dimensions of the objects.

It can be concluded that not only the electron energy, but also the properties of the object—density, chemical composition, and shape—have an impact on the distribution of absorbed dose throughout the object. Since biological objects have non-homogeneous density, complex shapes, and chemical composition, it is necessary to take an individual approach to establish the dose distribution for various objects.

4. Computer simulation of depth dose distributions

The non-uniformity of the absorbed dose distribution throughout the irradiated object entails the non-uniformity of the irradiation effect: overexposure can cause material destruction or undesirable changes of physical and chemical material properties, while underexposure may not lead to the desired effect [15–17]. Therefore, the distribution of the absorbed dose throughout the irradiated object should be strictly controlled.

Modern dosimetry methods for determining the integral dose absorbed by an object and its dose distribution during industrial irradiation are time-consuming and expensive [47, 48]. Dosimetry requires regular repetition and calibration due to possible changes in the electron beam spectrum during operation of the industrial accelerator [44, 49].

Computer simulation is highly accurate method that allows to calculate 2D and 3D depth dose distributions, taking into account all the physical and technical parameters

Code	Developer	Particle type	Particle energy	Material
EGSnrc [5]	National Research Council of Canada, Canada	electrons, photons	1 keV–10 GeV	Homogeneous materials; Simple material geometry (parallelepiped, cylinder, sphere)
PENELOPE [6]	Nuclear Energy Agency, France	electrons, photons	1 keV–1 GeV	
MCNPX [7]	Los Alamos National Laboratory, USA	electrons, photons neutrons	1 keV–100 GeV (photons) 1 keV–1 GeV (electrons)	Homogeneous and inhomogeneous materials; Complex material geometry
GEANT4 [8]	CERN, Switzerland	electrons, photons neutrons protons, muons	~10 eV–100 TeV	

Table 3.
Capabilities of transport modeling codes.

of the irradiation method, such as the geometry, density, and chemical composition of the irradiated object [50–55].

4.1 Monte Carlo radiation transport codes

The Monte Carlo method, which is used to establish transport codes, has proved to be efficient for simulating different types of irradiation for objects of various shapes and geometries [4, 52–55]. **Table 3** provides transport codes for simulating the interaction of particles with the matter as well as energy ranges and materials that can be used in simulation [56].

While the EGSnrc and PENELOPE codes are limited in their application to the objects of simple geometry [5, 6], transport codes MCNPX and GEANT4 are more widely used in the industry as these are more comprehensive instruments for simulating a wider variety of physical processes representing the interactions of electrons, photons, protons, and neutrons [7, 8]. The MCNPX code is used in nuclear medicine, radiation safety, accelerator development, and modeling industrial irradiation of biological objects and materials [52]. GEANT4 is currently the most complete set of tools for simulating the passage of particles through matter [8]. Its fields of application include high energy physics, nuclear and accelerator physics, and medical and space science research. Unlike other codes, GEANT4 simulates any geometry of the objects and the radiation source with any energy spectrum and traces physical processes selected for a particular irradiation method and objects, which makes GEANT4 the most flexible transport code with a wide scope of uses [57].

4.2 Computer simulation of the absorbed dose distribution in biological objects using GEANT4

During computer simulation biological objects irradiated with electron beams are usually represented by water phantoms since water is close in its properties to the range of biological objects involved in irradiation processing. To calculate the absorbed dose distribution using GEANT4, it is necessary to describe the geometry of the object and determine the radiation source and the volume for detecting the absorbed energy.

GEANT4 contains more than 20 standard shapes that can be used to simulate the geometry of biological objects. To increase the accuracy of depth dose distributions in biological objects, it is necessary to determine the geometry of the object as a set of volumes of different densities and chemical compositions that correspond to the different parts of the object. Electron beam is determined by specifying its size and shape, the coordinates of its center, electron energy spectrum, spatial distribution, and number of electrons emitted per second to simulate the irradiation method used. Detection is performed using virtual volumes, which divide the water phantom with the linear dimensions of $X \times Y \times Z$ into the number of $N_x \times N_y \times N_z$ cubic cells. In each cell, the total energy absorbed during interactions of electrons with matter is recorded using the following formula:

$$E_{i,j,k} = \sum_{n=0}^{N_{i,j,k}} \Delta \mathrm{E}_{i,j,k,n}, \tag{3}$$

where *i,j,k* is the cell index, $\Delta \mathrm{E}_{i,j,k,n}$ is the energy absorbed by the *i,j,k* cells during the n interaction, $N_{i,j,k}$ is the number of interactions occurring in the *i,j,k* cell.

The standard deviation of absorbed energy $S_{i,j,k}$ in the *i,j,k* cell is determined by the sum of the squares of the absorbed energy $\sum_{n=1}^{N_{i,j,k}} E_{i,j,k}^2$ and is calculated using the formula:

$$S_{i,j,k} = \sqrt{\frac{1}{N_{i,j,k}} \sum_{n=1}^{N_{i,j,k}} E_{i,j,k}^2 - \left(\frac{1}{N_{i,j,k}} \sum_{n=1}^{N_{i,j,k}} E_{i,j,k}\right)^2}. \tag{4}$$

Then, knowing the mass of the *i,j,k* cell $m_{i,j,k}$, the absorbed dose is determined by the formula

$$D_{i,j,k} = \frac{E_{i,j,k}}{m_{i,j,k}}. \tag{5}$$

To calculate a 3D-color map representing the relative absorbed dose distribution throughout the biological object the dose absorbed by each *i,j,k* cell of water phantom is color-coded, and each cell is marked with the color corresponding to the value $D_{i,j,k}^{rel}$, which is the ratio between $D_{i,j,k}$ and the maximum dose value in the water phantom.

5. Reconstruction of electron beam energy spectra

A necessary condition for successful irradiation with electron beams is complete information about the spatial distribution of the absorbed dose in the irradiated object, which is determined both by the properties of the irradiated object (i.e., geometry, elemental composition, and density) and by the source parameters, primarily the energy beam spectrum [13, 14]. Modern approaches to determine the energy spectra of accelerators are based on direct measurement of electron energy using special equipment [12] and on indirect methods based on the reconstruction of the spectra using experimentally measured data [9–11].

The indirect method of reconstructing the energy spectrum is based on the solution of the Fredholm integral equation of the first kind, which can be formulated as follows:

$$D(x) = \int_{0}^{E_{max}} f(E)d(x,E)dE, \quad (6)$$

where $D(x)$ is the distribution of some parameters, such as the absorbed charge, dose, fluence, and flux density, along the parameter x (depth, angle, etc.); $d(x,E)$ is the distribution of the same parameters for a monoenergetic beam with the energy E; $f(E)$ is the energy spectrum.

Usually, this equation is reduced to a system of linear algebraic equations by approximating a continuous spectrum $f(E)$ linear combination of basic functions $F_j(E)$ with a^j acting as weights and by approximating distribution $D(x)$ with a discrete set of values D^i at points x_i. The corresponding system of linear algebraic equations takes the form:

$$D^i = \sum_{j=1}^{N} d^{i,j} a^j. \quad (7)$$

Here $d^{i,j}$ is the parameter at the point x_i created by an electron beam with an energy spectrum $F_j(E)$.

A common method for solving the system (7) is the least squares method. Fredholm equation of the first kind in the general case is an incorrectly posed problem, that is, the solution of the equation may not exist or there may be several of them. Also, the solution of the system (7) can change greatly with small changes of $D(x)$. These properties of the integral equation are transferred to its discrete counterpart, which leads to non-physical sharply oscillating solutions that have little to do with the true spectrum of the beam, as demonstrated in **Figure 5a**.

To address this phenomenon, the regularization procedure is used [11, 58] that involves the modification of the original problem, which turns an incorrectly posed problem into a correctly posed one. There are two types of regularization: the first type modifies the equation; the second type modifies allowed solutions. The simplest

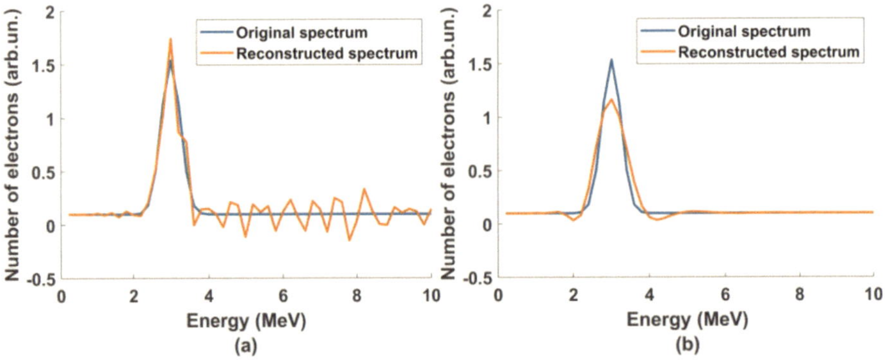

Figure 5.
(a) The appearance of non-physical oscillations in the reconstructed spectrum; (b) Peak smoothing due to over regularization.

example of regularization of the second type is the imposition of a non-negativity condition. Such regularization is used in problems in which the non-negativity of an unknown quantity is guaranteed by its physical nature, for example, mass, spectrum, energy, etc. The first method usually consists of the introduction of regularizing operators. One of the most popular methods is L_2 regularization, better known as Tikhonov regularization [58–60]:

$$L_2 = \sum_i \left(D^i - \sum_i d^{i,j} a^j\right)^2 + \alpha \bullet \theta\left[a^j\right], \tag{8}$$

where α is the regularization parameter, $\theta\left[a^j\right]$ is the regularizing operator.

There is a wide range of regularizing operators [61, 62] and values of regularization parameters. In general, the higher the value of the regularization parameter, the less the regularized problem will be related to the nonregularized problem, the smaller the value, the less noticeable the regularization effect will be. In the simplest case, the regularizing operator is a Euclidean norm L_2. From practice, it is known [63] that for the simplest regularizing operator, the best results are obtained by the residual method, which prescribes to choose such values of the regularization parameter that:

$$\sum_i \left(D^i - \sum_i d^{i,j} a^j\right)^2 \approx \sigma^2 \left\|D^i\right\|^2, \tag{9}$$

where σ is the relative error of D^i.

It is worth noting that the Tikhonov regularization with the simplest regularizing operator leads to a smoothing of the peaks of the spectrum (**Figure 5b**), which may be undesirable. This can be addressed by modifying the regularizing operator, for example [64]:

$$\theta\left[a^j\right] = \sum_j \log\left(a^j\right)^2. \tag{10}$$

Also, one can decompose the spectrum into a regular and singular part [64]:

$$f(E) = f_s(E) + f_r(E), \tag{11}$$

where $f_r(E)$ is regular and $f_s(E)$ is singular component:

$$f_s(E) = \beta e^{-\frac{(E-\mu)^2}{\Delta^2}}. \tag{12}$$

Such decomposition allows to treat two parts of spectrum separately and can lead to improved results.

6. Methods to increase dose uniformity in the objects irradiated with accelerated electrons

Sterilization of biological and non-biological objects using accelerated electrons involves the reduction of microbial growth to the required level due to high energy values absorbed by microorganisms. As electrons penetrate biological objects, they lose energy in direct interaction with atomic electrons of bacteria cell structures, which results in DNA breaks and the destruction of cell membrane [1, 3]. Reactive oxygen species appearing during radiolysis of water in biological objects destroy chemical bonds in cell molecules causing inactivation of microorganisms. Moreover, the higher the dose absorbed by the object, the greater number of microorganisms inhibited during irradiation. However, the increase in irradiation dose is limited by the irreversible physical and chemical changes occurring in the biological object. Considering that each object has its specific chemical composition and physical properties, it is necessary to select the dose range individually to ensure that effective irradiation dose does not lead to irreversible changes in the object. The dose range for the treatment of biological objects is narrower compared to non-biological objects due to a great complexity of biochemical composition of such objects, and even a small change in the dose can lead to a significant alteration in the structure and functionality of cells.

Treatment of biological objects with accelerated electrons is characterized by the non-uniformity of dose distribution over the volume of irradiated objects due to the nature of dose distribution, non-homogeneous density of biological objects, complex geometry, and chemical composition. The distribution of radiation effect on both microbiological parameters and properties of the object is non-homogeneous, which reduces the efficiency of irradiation treatment and makes it difficult to maintain the microbial values throughout the object at the required level. Thus, it is highly important to develop feasible methods to increase uniformity of absorbed dose distribution over the volume of irradiated objects.

Currently, the following methods are used in industrial irradiation facilities to increase the uniformity of dose distribution:

- Two-side irradiation for effective treatment of the objects with a high thickness;
- Application of special polymer absorbers with a density close to that of the irradiated object to fill the empty space in the package, making the complex shape of the object close to a parallelepiped;
- Varying the energy of accelerated electrons in several irradiation sessions.

When the thickness of the object exceeds the maximum path of electrons with energies up to 10 MeV, it is reasonable to carry out double-side irradiation [17]. However, this is not applicable to objects with a shape different from that of a parallelepiped because it does not provide consistent irradiation uniformity.

The second way to improve the uniformity of irradiation is the use of polymer absorbers designed to fill the empty space in the package or irregularities of the irradiated object, imitating the shape of a parallelepiped to enable the use of the previous method [17]. However, this way of tackling the problem of irradiation non-uniformity is not cost-effective since it requires the replacement of polymers that are destroyed during irradiation.

Another way to increase the irradiation uniformity is to vary the energy of accelerated electrons during several irradiation sessions. **Figure 6** shows the dependencies of the absorbed dose on the depth of a parallelepiped irradiated with 3 MeV and 10 MeV electrons, as well as the distribution of the absorbed dose when the

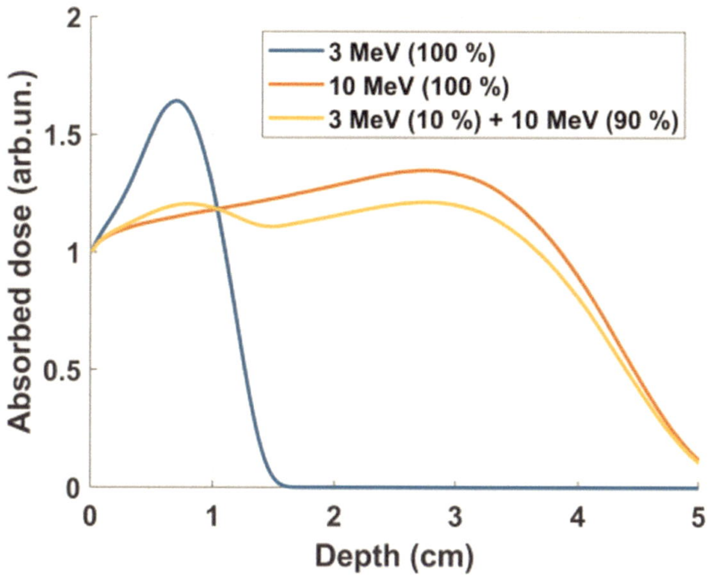

Figure 6.
Distribution of absorbed dose in the water parallelepiped during irradiation with 3 MeV (blue line) and 10 MeV (red line) electrons, as well as irradiation with the combination of 3 MeV and 10 MeV electron beams (yellow line).

parallelepiped is irradiated by a combination of 3 MeV and 10 MeV electron, beams with weighted coefficients 0.1 and 0.9, respectively.

As can be seen from **Figure 6**, the combination of two irradiation energies allows to increase the dose uniformity. However, the use of multiple sessions increases treatment time, which makes it difficult to apply to a wide range of categories of biological objects.

6.1 Electron beam modification method for improving dose uniformity

This study proposes a method of modifying the beam spectrum using aluminum plates, which allows to increase the dose uniformity during irradiation with accelerated electrons with the energy up to 10 MeV [44]. The main idea of increasing the uniformity of irradiation consists in additional placement of modifier plates between the beam output and the irradiated object. The energy and angular distributions of the directed monoenergetic electron beam are blurred after passing through the plate, which leads to the appearance of electrons with lower energies in the beam, improving the dose uniformity throughout the object during one irradiation session.

Figure 7 shows that the addition of aluminum modifier plates changes the dose distribution throughout the irradiated object. The dose value increases in the surface layers at a distance ranging from 0 to 1.5 g/cm^2, while the maximum electron path in the substance decreases. **Figure 8** shows that the uniformity coefficient K grows linearly with increasing thickness of the aluminum modifier plate d.

For mass thicknesses of the object L ranging from 1.025 to 3.125 g/cm^2 with an error of no more than 5% it is possible to select the thickness of the aluminum modifier plate d in the range from 0.5 to 5 mm, at which the thickness of the object

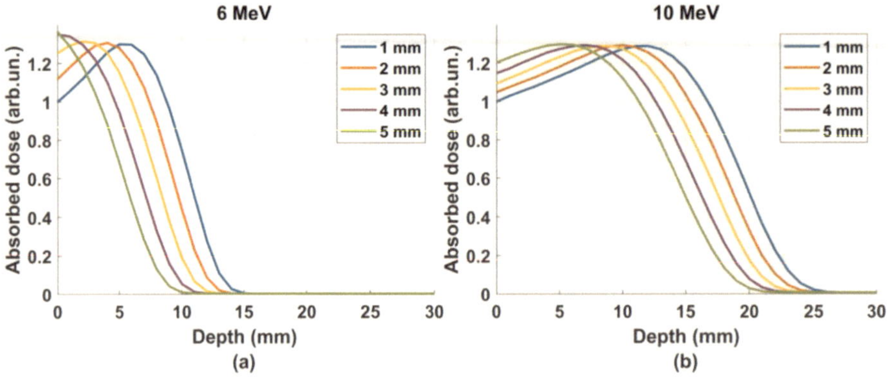

Figure 7.
Dependency of the relative absorbed dose on the depth in water phantom irradiated with 6 MeV (a) and 10 MeV (b) electron beams with the aluminum modifier plates with thicknesses ranging from 1 mm to 5 mm.

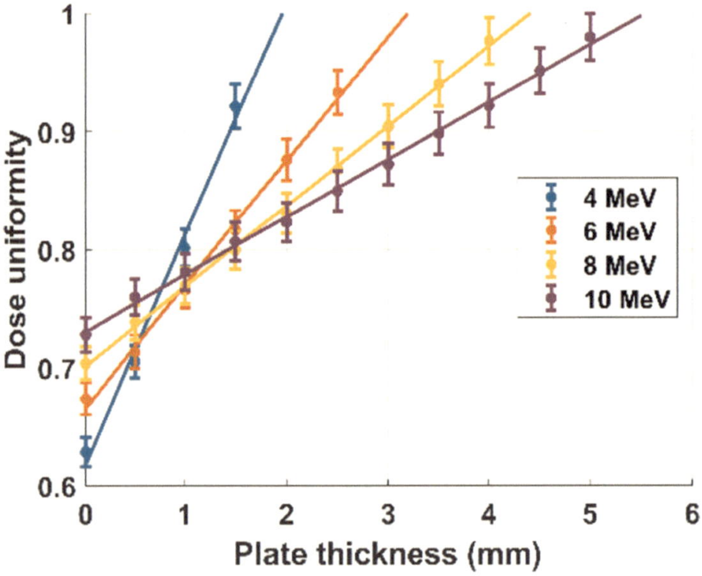

Figure 8.
Dependency of irradiation uniformity K *of a water parallelepiped on the thickness of the aluminum modifier plate after irradiation with 4 MeV, 6 MeV, 8 MeV, and 10 MeV electron beams [44].*

corresponds to the optimal distance when the object is irradiated with electron beams with the energy ranging from 4 MeV to 10 MeV using the following formula:

$$d[\text{cm}] = -0.060[\text{cm}] - 0.199\left[\frac{\text{cm}^3}{\text{g}}\right] \times L\left[\frac{\text{g}}{\text{cm}^2}\right] + 0.093\left[\frac{\text{cm}}{\text{MeV}}\right] \times E_0[\text{MeV}] \\ - - 0.002\left[\frac{\text{cm}^3}{\text{MeV}\bullet\text{g}}\right] \times E_0[\text{MeV}] \times L\left[\frac{\text{g}}{\text{cm}^2}\right], \quad (13)$$

and to calculate with an error not exceeding 5% the irradiation uniformity K for different combinations of modifier thicknesses d in the range from 0.5 mm to 5 mm and initial electron energies E_0:

$$K = 0.603 - 1.830\left[\frac{1}{\text{cm}}\right] \times d[\text{cm}] + 0.012\left[\frac{1}{\text{MeV}}\right] \times E_0[\text{MeV}] - -0.135\left[\frac{1}{\text{MeV}\bullet\text{cm}}\right] \times E_0[\text{MeV}] \times d[\text{cm}]. \tag{14}$$

Thus, knowing the required minimum value of the coefficient K_{min}, it is possible to select combinations of electron beam energies and thicknesses of modifier plates, at which for a parallelepiped-shaped object the irradiation uniformity is achieved at a level that is not less than the required $K \geq K_{min}$ over its volume.

The proposed method allows to increase the irradiation uniformity up to 0.97 in parallelepiped-shaped objects irradiated with electron beam energy between 4 MeV and 10 MeV when using aluminum modifier plates with a thickness ranging from 0.5 mm to 5.5 mm. The optimal distance from the surface of the object at which the dose value is equal to the surface dose ranges from 1.025 g/cm^2 to 3.125 g/cm^2 and decreases with an increase in modifier plate thickness [65].

6.2 Linear combination of modifier plates for improved dose uniformity

It is possible to increase maximum thickness of the irradiated objects while maintaining the dose uniformity at a high level by using the combination of modifier plates of different thicknesses in such a way that the absorbed dose distribution in the object is as close as possible to a constant value, or, equivalently, and by minimizing the following function to calculate the weight coefficients a^i:

$$\sum_{i=1}^{N}\left(\sum_{j=1}^{M} a^i d^{i,j} - const\right)^2 \rightarrow Min, \tag{15}$$

where $d^{i,j}$ is the dose at the depth x_j of the object irradiated with electrons by adding aluminum modifier plate with a thickness of d_i. Summation is carried out over i ranging from 1 to N, where N is the number of aluminum plates of different thicknesses, and over j ranging from 1 to M, where M is the number of points in the object at which the absorbed dose is determined.

Figure 9a shows the absorbed dose distribution in the water phantom irradiated with 10 MeV electrons without modifier plates, with one plate, and with a combination of plates with the thickness varying from 0.1 mm to 9.5 mm. **Figure 9b** shows the absorbed dose distributions in the water phantom irradiated with electron beams with the energy ranging from 5 MeV to 10 MeV by adding a combination of modifier plates. It can be seen while the phantom irradiated without adding plates has an optimal distance 3.8 g/cm^2 and the dose uniformity 0.71, adding one modifier plate increases the dose uniformity to 0.97 and decreases the optimal distance to 1.2 g/cm^2. In contrast, using the combination of plates maintains the optimal distance at 3 g/cm^2 and the dose uniformity at 0.98 (**Figure 9a**). **Figure 9b** shows that varying the electron energy and using a combination of modifier plates allows to increase the dose uniformity up to 0.95 for objects with the mass thickness ranging from 1.8 g/cm^2 to 3.8 g/cm^2.

It should be noted that the proposed algorithm makes it possible to select the combination of plates that results in maximum dose uniformity for any beam spectrum. The maximum possible dose uniformity of 0.98 is achieved for objects with a mass thickness of up to 3.5 g/cm^2 irradiated with 10 MeV electrons.

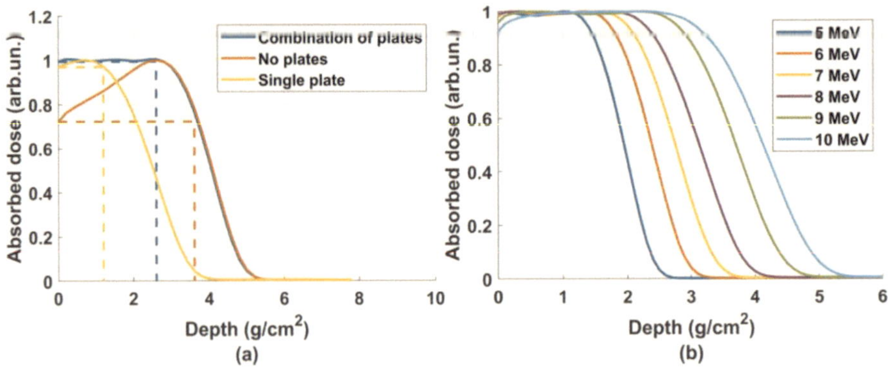

Figure 9.
(a) Absorbed dose distribution during 10 MeV irradiation with and without plates; (b) Absorbed dose distributions in the water phantom during irradiation with 5–10 MeV electrons with the combination plates [66].

7. Conclusion

The analysis of various applications of electron beams shows that the dose ranges, electron energy, and beam power vary greatly depending on the purpose of irradiation, which determines the electron penetration depth, absorbed dose distribution, and the desired effect. The study uses the GEANT4 toolkit to simulate the impact of electron beam irradiation on the absorbed dose distribution depending on electron energy, density, chemical composition, and geometry of the irradiated object.

To increase the accuracy of computer simulation in the absence of open access to energy beam spectrum of industrial accelerators our team developed an algorithm to reconstruct the electron beam spectrum which would allow to calculate depth dose distribution in an object of any geometry and chemical composition with an accuracy of up to 95%. Using our extensive collection of GEANT4 data on irradiation of biological objects with electrons having the energy of up to 10 MeV we developed a method to increase the irradiation uniformity up to 0.98 for the objects with a mass thickness of up to 3.5 g/cm^2 when one-side irradiated with 10 MeV electrons.

Since computer simulation found that the edges of the object with the shape close to a parallelepiped are underexposed, our further research will be aimed at increasing irradiation uniformity throughout the object irradiated with electron beams. Another area of interest for research is to create a database of absorbed dose distribution for irradiation of biological objects with photons to develop an algorithm for reconstructing bremsstrahlung irradiation energy spectra, which is commonly used in irradiation facilities.

Acknowledgements

We thank researchers at Scobeltsyn Scientific Research Institute of Nuclear Physics, Lomonosov Moscow State University for conducting experiments using electron accelerator.

This research was funded by the Russian Science Foundation, grant number 22-63-00075.

Electron Beam Processing of Biological Objects and Materials
DOI: http://dx.doi.org/10.5772/intechopen.112699

Author details

Ulyana Bliznyuk*, Aleksandr Chernyaev, Victoria Ipatova, Aleksandr Nikitchenko, Felix Studenikin and Sergei Zolotov
Lomonosov Moscow State University, Moscow, Russia

*Address all correspondence to: uabliznyuk@gmail.com

References

[1] International Atomic Energy Agency. Trends in Radiation Sterilization of Health Care Products. Vienna: International Atomic Energy Agency; 2008

[2] Dorado F. Atoms Radiation and Radiation Protection. 3rd ed. Weinheim: Wiley-VCH; 2007. p. 595

[3] Venturi M, D'Angelantonio M, editors. Applications of radiation chemistry in the fields of industry, biotechnology and environment. In: Topics in Current Chemistry Collections. Switzerland: Springer International Publishing; 2017. p. 309. DOI: 10.1007/978-3-319-54145-7

[4] International Atomic Energy Agency. IAEA Radiation Technology Series No. 1. Use of Mathematical Modelling in Electron Beam Processing: A Guidebook. Vienna: International Atomic Energy Agency; 2010

[5] Kawrakow I. The EGSnrc Code System: Monte Carlo Simulation of Electron and Photon Transport. Manual – Guides. Canada: NRC; 2023. p. 323

[6] Nuclear Energy Agency. PENELOPE 2018: A code system for Monte Carlo simulation of electron and photon transport. In: Workshop Proceedings; 28 January–1 February 2019. Barcelona, Spain; 2019. p. 420. DOI: 10.1787/32da5043-en

[7] The MCNP® Code [Internet]. 2023. Available from: https://mcnp.lanl.gov [Accessed: 23 July 2023]

[8] Allison J, Amako K, Apostolakis J, Arce P, Asai M, Aso T, et al. Recent developments in Geant 4. Nuclear Instruments and Methods in Physics Research Section A: Accelerators, Spectrometers, Detectors and Associated Equipment. 2016;**835**:186-225. DOI: 10.1016/j.nima.2016.06.125

[9] Torres-Díaz J, Grad GB, Bonzi EV. Measurement of linear accelerator spectra, reconstructed from percentage depth dose curves by neural networks. Physica Medica: European Journal of Medical Physics. 2022;**96**:81-89. DOI: 10.1016/j.ejmp.2022.02.019

[10] Bliznyuk UA, Avdyukhina VM, Borshchegovskaya PY, Ipatova VS, Nikitchenko AD, Studenikin FR, et al. Estimating the accuracy of reconstructing Bichromatic spectra of electron beams from depth dose distributions. Bulletin of the Russian Academy of Sciences: Physics. 2021; **85**(10):1108-1112. DOI: 10.3103/S1062873821100099

[11] Visbal JHW, Costa AM. Inverse reconstruction of energy spectra of clinical electron beams using the generalized simulated annealing method. Radiation Physics and Chemistry. 2019; **162**:31-38. DOI: 10.1016/j.radphyschem.2019.04.022

[12] Kozyrev AV, Kozhevnikov VY, Vorobyov MS, Baksht EK, Burachenko AG, Koval NN, et al. Reconstruction of electron beam energy spectra for vacuum and gas diodes. Laser and Particle Beams. 2015;**33**(02):183-192. DOI: 10.1017/s0263034615000324

[13] Li G, Wu A, Lin H, Wu Y. Electron spectrum reconstruction as nonlinear programming model using micro-adjusting algorithm. In: Peng Y, Weng X, editors. 7th Asian-Pacific Conference on Medical and Biological Engineering. IFMBE Proceedings. Berlin, Heidelberg: Springer; 2008. p. 19. DOI: 10.1007/978-3-540-79039-6_112

[14] Gui L, Hui L, Ai-Dong W, Gang S, Yi-Can W. Realization and comparison of several regression algorithms for Electron energy Spectrum reconstruction. Chinese Physics Letters. 2008;**25**(7):2710. DOI: 10.1088/0256-307X/25/7/104

[15] International Atomic Energy Agency. Development of electron Beam and x Ray Applications for Food Irradiation. Vienna: International Atomic Energy Agency; 2022. p. 372

[16] International Atomic Energy Agency. Manual of Good Practice in Food Irradiation : Sanitary, Phytosanitary and Other Applications. Vienna: International Atomic Energy Agency; 2015. p. 104

[17] Miller RB. Electronic Irradiation of Foods: An Introduction to the Technology. Springer - Food Engineering Series. 1st ed. New York: Springer; 2005. p. 296. DOI: 10.1007/0-387-28386-2

[18] Ozer ZN. Electron beam irradiation processing for industrial and medical applications. EPJ Web of Conferences. 2017;**154**:01019. DOI: 10.1051/epjconf/201715401019

[19] Miriam S, Unnati P. Electron beam irradiation-an environmenttally safe method of fungal decontamination and food preservation: A review. International Journal of Life Sciences. 2015;**A5**:7-10

[20] Shvedunov VI, Alimov AS, Ermakov AN, Kamanin AN, Khankin VV, Kurilik AS, et al. Electron accelerators design and construction at Lomonosov Moscow State University. Radiation Physics and Chemistry. 2019;**159**:95-100. DOI: 10.1016/j.radphyschem.2019.02.044

[21] Abou Elmaaty T, Okubayashi S, Elsisi H, Abouelenin S. Electron beam irradiation treatment of textiles materials: A review. Journal of Polymer Research. 2022;**29**:117. DOI: 10.1007/s10965-022-02952-4

[22] Cleland MR. Industrial applications of electron accelerators. 2006;**2006**:383-416. DOI: 10.5170/CERN-2006-012.383

[23] Sun Y, Chmielewski AG, editors. Applications of Ionizing Radiation in Materials Processing. Warszaww: Institute of Nuclear Chemistry and Technology; 2017. pp. 56-80

[24] Alimov S. Practical applications of electron accelerators. Preprint SINP MSU, Moscow. 2011. 41 p

[25] Sabharwal S. Electron beam irradiation applications. In: Proceedings of PAC2013. Pasadena, CA, USA; 2013. pp. 745-748

[26] Gong X, Anderson T, Chou K. Review on powder-based electron beam additive manufacturing technology. Manufacturing Review. 2014;**1**(2):1-12. DOI: 10.1051/mfreview/2014001

[27] Chaudhary S, Avinashi SK, Rao J, Gautam C. Recent advances in additive manufacturing, applications and challenges for dentistry: A review. ACS Biomaterials Science & Engineering. 2023;**9**(7):3987-4019. DOI: 10.1021/acsbiomaterials.2c01561

[28] Manaila E, Craciun G, Ighigeanu D, Lungu IB, Dumitru Grivei MD, Stelescu MD. Degradation by Electron beam irradiation of some composites based on natural rubber reinforced with mineral and organic fillers. International Journal of Molecular Sciences. 2022;**23**(13):6925. DOI: 10.3390/ijms23136925

[29] Working Material Produced by the International Atomic Energy Agency. In: Radiation Effects on Polymer Materials Commonly Used in Medical Devices. Report of the 1st RCM for CRP F2303; 15–19 November 2021; Vienna, Austria; 2021. 170 p

[30] Shahidi S. Effect of irradiation for producing the conductive and smart hydrogels. Cellulose-Based Superabsorbent Hydrogels: Polymers and Polymeric Composites: A Reference Series. 2019;**2019**:625-653. DOI: 10.1007/978-3-319-77830-3_22

[31] Ashfaq A, Clochard M-C, Coqueret X, Dispenza C, Driscoll MS, Ulański P, et al. Polymerization reactions and modifications of polymers by ionizing radiation. Polymers. 2020; **12**(12):2877. DOI: 10.3390/polym12122877

[32] Vega-Hernández MÁ, Cano-Díaz GS, Vivaldo-Lima E, Rosas-Aburto A, Hernández-Luna MG, Martinez A, et al. A review on the synthesis, characterization, and Modeling of polymer grafting. PRO. 2021;**9**(2):375. DOI: 10.3390/pr9020375

[33] Berejka AJ, Kaluska IM. Materials used in medical devices. In: Trends in Radiation Sterilization of Health Care Products. Vienna: International Atomic Energy Agency; 2008. pp. 119-128

[34] Mrazova H, Koller J, Kubisova K, Fujerikova G, Klincova E, Babal P. Comparison of structural changes in skin and amnion tissue grafts for transplantation induced by gamma and electron beam irradiation for sterilization. Cell and Tissue Banking. 2016;**17**:255-260. DOI: 10.1007/s10561-015-9536-3

[35] Clemmons HE, Clemmons EJ, Brown EJ. Electron beam processing technology for food processing. Electron Beam Pasteurization and Complementary Food Processing Technologies. 2015;**2015**:11-25. DOI: 10.1533/9781782421085.1.11

[36] Shayanfar S, Pillai SD. Future trends in electron beam technology for food processing. Electron Beam Pasteurization and Complementary Food Processing Technologies. 2015;**2015**: 295-309. DOI: 10.1533/9781782421085.3.295

[37] Chulikova N, Malyuga A, Borshchegovskaya P, Zubritskaya Y, Ipatova V, Chernyaev A, et al. Electron beam irradiation to control *Rhizoctonia solani* in potato. Agriculture. 2023;**13**: 1221. DOI: 10.3390/agriculture13061221

[38] Bliznyuk U, Avdyukhina V, Borshchegovskaya P, Bolotnik T, Ipatova V, Nikitina Z, et al. Effect of electron and X-ray irradiation on microbiological and chemical parameters of chilled Turkey. Scientific Reports. 2022;**12**:750. DOI: 10.1038/s41598-021-04733-3

[39] Hossain K, Maruthi YA, Das NL, Rawat KP, Sarma KSS. Irradiation of wastewater with electron beam is a key to sustainable smart/green cities: A review. Applied Water Science. 2018;**8**(1):1-11. DOI: 10.1007/s13201-018-0645-6

[40] Emami-Meibodi M, Parsaeian MR, Amraei R, Banaei M, Anvari F, Tahami SMR, et al. An experimental investigation of wastewater treatment using electron beam irradiation. Radiation Physics and Chemistry. 2016; **125**:82-87. DOI: 10.1016/j.radphyschem.2016.03.011

[41] Ezzeldien M, Amer MI, Shalaby MS, Moustaf SH, Hashem HM, Emam-Ismail M, et al. Electron beam irradiation-induced changes in the

microstructure and optoelectronic properties of nanostructured Co-doped SnO_2 diluted magnetic semiconductor thin film. European Physical Journal Plus. 2022;**137**(8):905. DOI: 10.1140/epjp/s13360-022-03079-7

[42] Noithong P, Pakkong P, Naemchanthara K. Color change of Spodumene gemstone by Electron beam irradiation. Advanced Materials Research. 2013;**770**:370-373. DOI: 10.4028/www.scientific.net/amr.770.370

[43] Rudychev VG, Azarenkov MO, Girka IO, Lazurik VT, Rudychev YV. Optimization of converter and bremsstrahlung characteristics for object irradiation. Radiation Physics and Chemistry. 2023;**206**:110815. DOI: 10.1016/j.radphyschem.2023.110815

[44] Studenikin FR, Bliznyuk UA, Chernyaev AP, Krusanov GA, Nikitchenko AD, Zolotov SA, et al. Electron beam modification for improving dose uniformity in irradiated objects. The European Physical Journal Special Topics. 2023;**232**:1631-1635. DOI: 10.1140/epjs/s11734-023-00886-6

[45] Bliznyuk UA, Studenikin FR, Borshchegovskaya PY, Krusanov GA, Ipatova VS, Chernyaev AP. Characteristics of dose distributions of Electron beams used in the radiation processing of food products. Bulletin of the Russian Academy of Sciences: Physics. 2021;**85**(10):1097-1101. DOI: 10.3103/S1062873821100087

[46] El-Ghossain MO. Calculations of stopping power, and range of electrons interaction with different material and human body parts. International Journal of Science and Technology Research. 2017;**6**(1):114-118

[47] Helt-Hansen J, Miller A, Sharpe P, Laurell B, Weiss D, Pageau G. Dµ—A new concept in industrial low-energy electron dosimetry. Radiation Physics and Chemistry. 2010;**79**(1):66-74. DOI: 10.1016/j.radphyschem.2009.09.002

[48] Devic S. Radiochromic film dosimetry: Past, present, and future. Physica Medica. 2011;**27**(3):122-134. DOI: 10.1016/j.ejmp.2010.10.001

[49] ISO/ASTM 51261. Practice for calibration of routine dosimetry systems for radiation processing. 2013

[50] Nordlund K. Historical review of computer simulation of radiation effects in materials. Journal of Nuclear Materials. 2019;**520**:273e295. DOI: 10.1016/j.jnucmat.2019.04.028

[51] Connaghan JP, Saylor MC, Calvert GW, Yeadon SC, Pyne CH, Mellor P, et al. Mathematical modeling of industrial radiation processes application and end-user training. Radiation Physics and Chemistry. 2004;**71**(1–2):335-338. DOI: 10.1016/j.radphyschem.2004.04.011

[52] Peivaste I, Alahyarizadeh G. Comparative study on absorbed dose distribution of potato and onion in X-ray and Electron beam system by MCNPX2.6 code. Mapan. 2019;**34**:19-29. DOI: 10.1007/s12647-018-0287-z

[53] Qin H, Yang G, Kuang S, Wang Q, Liu J, Zhang X, et al. Concept development of X-ray mass thickness detection for irradiated items upon electron beam irradiation processing. Radiation Physics and Chemistry. 2018;**143**:8-13. DOI: 10.1016/j.radphyschem.2017.09.012

[54] Cárcel JA, Benedito J, Cambero MI, Cabeza MC, Ordóñez JA. Modeling and

optimization of the E-beam treatment of chicken steaks and hamburgers, considering food safety, shelf-life, and sensory quality. Food and Bioproducts Processing. 2015;**96**:133-144. DOI: 10.1016/j.fbp.2015.07.006

[55] Kim J, Moreira RG, Castell-Perez ME. Validation of irradiation of broccoli with a 10MeV electron beam accelerator. Journal of Food Engineering. 2008; **86**(4):595-603. DOI: 10.1016/j.jfoodeng.2007.11.018

[56] The Panel on Gamma and Electron Irradiation Modelling Working Group. Review of Monte Carlo Modelling Codes. London; 2007. p. 26

[57] Reed RA, Weller RA, Akkerman A, Barak J, Culpepper W, Duzellier S, et al. Anthology of the development of radiation transport tools as applied to single event effects. IEEE Transactions on Nuclear Science. 2013;**60**(3): 1876-1911. DOI: 10.1109/tns.2013.2262101

[58] Wazwaz A-M. The regularization method for Fredholm integral equations of the first kind. Computers & Mathematics with Applications. 2011; **61**(10):2981-2986. DOI: 10.1016/j.camwa.2011.03.083

[59] Xu Y, Pei Y, Dong F. An adaptive Tikhonov regularization parameter choice method for electrical resistance tomography. Flow Measurement and Instrumentation. 2016;**50**:1-12. DOI: 10.1016/j.flowmeasinst.2016.05.004

[60] Tikhonov AN. Ill-posed problems of linear algebra and a stable method for their solution. Doklady Akademii Nauk SSSR. 1965;**163**(3):591-594

[61] Esuabana IM, Abasiekwere UA. A survey of regularization methods of solution of Volterra integral equations of the first kind. Applied Mathematics. 2018;**8**(3):33-41. DOI: 10.5923/j.am.20180803.01

[62] Leonov AS. On quasioptimum selection of the regularization parameter in M. M. Lavrent'ev's method. Siberian Mathematical Journal. 1993;**34**(4): 695-703. DOI: 10.1007/bf00975172

[63] Bliznyuk UA, Borshchegovskaya PY, Ipatova VS, Nikitchenko AD, Studenikin FR, Chernyaev AP. Determining the beam Spectrum of industrial Electron accelerator using depth dose distribution. Bulletin of the Russian Academy of Sciences Physics. 2022;**86**:500-507. DOI: 10.3103/S1062873822040062

[64] Wei J, Sandison GA, Chvetsov AV. Reconstruction of electron spectra from depth doses with adaptive regularization. Medical Physics. 2006; **33**(2):354-359. DOI: 10.1118/1.2161404

[65] Studenikin FR, Bliznyuk UA, Chernyaev AP, Khankin VV, Krusanov GA. Impact of Aluminum plates on uniformity of depth dose distribution in object during Electron processing. Moscow University Physics Bulletin. 2021;**76**(1):S1-S7. DOI: 10.3103/S0027134922010106

[66] Bliznyuk UA, Borshchegovskaya PY, Zolotov SA, Ipatova VS, Krusanov GA, Nikitchenko AD, et al. Reconstruction of depth dose distributions in materials created by electron beam. Physics of Particles and Nuclei. 2023;**54**(4): 575-580. DOI: 10.1134/S1063779623040081

Chapter 6

DC Parallel Ribbon Ion Beams for High-Dose Processes

Nicholas R. White

Abstract

Ribbon beams of heavy ions have advantages over cylindrical beams, including higher space-charge limits. History of use goes from Calutrons, Freeman and Bernas ion sources, to the first ion implanters in the 1970s. In the 1990s, 300 mm uniform parallel mass-analyzed ribbon beams were developed to enable precise doping by mechanically scanning a substrate through the ion beam in one dimension. Ion species included the primary dopants boron, phosphorus and arsenic, but many others are also used. Such sources can produce currents of heavy ions with linear current densities at the source of the order of 10 mA/cm, but these sources are limited in the beam breadth they can produce. Broader beams are used for flat-panel display manufacture. A new linear source design combines a modified Penning trap with magnetic cusp confinement system, allowing extension of linear sources to meter scale beam breadths, maintaining around 10 mA/cm linear current density. Magnetic analysis of such beams has required new developments because the weight of conventional dipole magnets increases very steeply as the pole gap is increased. A new magnetic configuration has been developed to address this issue, reducing the potential weight of meter-scale analyzed systems by an order of magnitude.

Keywords: ribbon beams, 300 mm, ion implantation, flat panel doping, scaling laws, uniformity, parallelism, purity, analyzing magnets, space-charge neutralization, instability

1. Introduction

Ion implantation of semiconductors is the largest application of ion beam processing. In integrated circuits, flat-panel displays, and other devices, precise regions are implanted through masks with n-type and p-type dopants and other materials. Additional applications of ion beams exist, typically requiring high currents and fairly precise control of the ion dose. Ribbon beams are also still used in isotope separator systems, where pure isotopes are required for medical and other purposes. This chapter focusses on parallel DC ribbon ion beams, i.e. beams in which the breadth greatly exceeds its thickness; in this work I shall use the term beam breadth to indicate the major dimension, and thickness to indicate the minor dimension, regardless of system orientation in space. This is consistent with the ribbon analogy.

An excellent review of ion beam technology through its first century was given by Freeman [1]. The first isotope separators were developed by E.O. Lawrence based on his experiments in the 37-inch cyclotron at Berkeley, and used for uranium separation in the Manhattan project in the 1940s [2], using 16-inch broad ribbon beams. The ion beam was generated in an ion source immersed in the same uniform magnetic field used to separate the isotopes; there was therefore a strong and uniform magnetic field in the ion source oriented in the direction of the beam breadth. Electrons from a filament, inside a chamber with a single slot-shaped exit, were trapped by this magnetic field in a tight cylindrical region, where they ionized the uranium-containing vapor, from which the ribbon beam was extracted.

In the 1950s Freeman worked on applying aspects of this technology to peaceful purposes. His ion source used a magnetic field generated by a dedicated magnet which was separate from the analyzer magnet used for isotope separation. He ran a straight tungsten filament down the center of the region where it was desired to ionize the ion source vapor, and electrons were confined in complex pattens is a zone surrounding this filament. René Bernas [3] devised a similar ion source, with a coiled filament at one end of a similar arc chamber.

Freeman at Harwell then collaborated with Lintott Engineering in Horsham, UK, represented in the US by High Voltage Engineering in Massachusetts, to develop ion beam systems operating at tens of keV with mass resolving power around 60–100 to create pure beams of dopants for ion implantation [4]. Lintott's early implanters were not known for reliability. Implanters were then commercialized by several companies in the 1970s including Extrion Corp, founded by Rose and others. The first commercially successful implanters tended to use lower currents than 1 mA, and scanned their ion beams across the surface of small silicon wafers; while usable, this scanning introduced variations in the incident angle, which resulted in variations in the range of the implanted ions, since the ions at certain incident directions could undergo planar or axial channeling in the monocrystalline silicon substrates. These systems often used a cold-cathode Penning ion source, producing a cylindrical beam of a few hundred microamps.

In 1988 Peter Rose founded Nova Associates Inc., with Ryding and Wittkower, to develop high-current implanters, starting with the Nova NV10–80. These took full account of the need for space-charge neutralization to transport currents of 10 mA or more of heavy ions at energies from 10 to 80 keV through a mass analyzer. These used a Freeman ion source, generating a ribbon beam about 44 mm tall and 3 mm thick at the ion source exit. Since a space-charge neutralized beam cannot be scanned electrostatically, they used two-dimensional mechanical scanning to uniformly raster the beam across the wafer surface. A batch of wafers was loaded into a circle onto a disk, which was spun to provide one direction of motion, and translated slowly in a radial direction to provide the other.

2. Limits to high-current beam extraction

The maximum current that can be extracted in an ion beam is determined by the space-charge limit, which for a cylindrical beam is approximately:

$$\Pi = I\sqrt{\frac{M}{2q}} = \left(\frac{4\varepsilon_0}{9}\right) V^{3/2} \left(\frac{r^2}{g^2}\right) \qquad (1)$$

where I is the ion current, M the ion mass, q the ion charge, ε_0 is the permittivity of free space, V the ion beam extraction voltage, g the gap across which this voltage is applied, and r the beam radius. The equation is exact for a beam originating on one plane surface and being accelerated to a second plane surface at a potential V. Since the beam passes through a hole in the second electrode, the potential at the center of the hole is less than V, so in in practice the beam radius r cannot exceed g. In this case, the equation is not exact, since r/g must be <1 in practice, and so it provides an upper limit to the maximum current. The quantity Π is usually called the beam poissance when used for heavy ions. However, for a ribbon beam this equation becomes:

$$\Pi = I\sqrt{\frac{M}{2q}} = \left(\frac{4\varepsilon_0}{9}\right)V^{3/2}\left(\frac{ab}{g^2}\right) \quad (2)$$

where a is the beam thickness and b the ribbon breadth at the source aperture. Now the maximum is set by the condition $a/g < 1$. It is now possible for b to greatly exceed a without the potential at the center of the electrode being significantly different from V, and as a result the perveance from a ribbon-shaped ion beam system can exceed that from a cylindrical system by a factor of the order of b/a. Thus a ribbon beam system in the case of the Nova NV-10 high-current system can carry ~44/3 times more current than a cylindrical beam.

3. Space-charge neutralization

A high-current ion beam contains ions which mutually repel each other, and a beam of the maximum current will diverge very rapidly, unless steps are taken to control this space-charge blowup. This process was known from the first days of the Calutron; the blowup can be mitigated by trapping electrons in the positively charged ion beam so that the space-charge of the ions is largely counteracted by the electrons. However, for a few decades the details of this process were not widely understood. Bernas [3] showed that the necessary electrons could be generated by collisions with residual gas atoms in several tens of microseconds, depending on the pressure and other conditions. It was understood since the 1960s that the extraction electrodes for a high-current ion beam have to be a triode, as in **Figure 1**, with the intermediate electrode at a more negative potential than the final beam potential, and of course the ion source at a positive potential, defining the final beam energy. The negative electrode, known as the suppression electrode, should repel electrons that find their way into the ion beam, and prevent them from being accelerated toward the ion source.

Ideally the electrons trapped in the positive ion beam will reach thermal equilibrium through multiple elastic collisions, and an unusual plasma will form, comprising the fast beam ions, a component of ~room-temperature positive ions from the residual gas, mainly generated through charge-exchange processes with the beam, and a population of electrons in thermal equilibrium, the density of which is roughly equal to the total ion density. Hiroyuki Ito, in his Ph. D. thesis [5], has modeled this population and shown that the positive ions typically occupy a zone in which the potential varies by 0.5 kT_e, where kT_e, the electron temperature, can be measured and is typically between about 3 and 8 eV in a beam 100% contained in conductive grounded walls, except for a negative suppression electrode as shown in **Figure 1**.

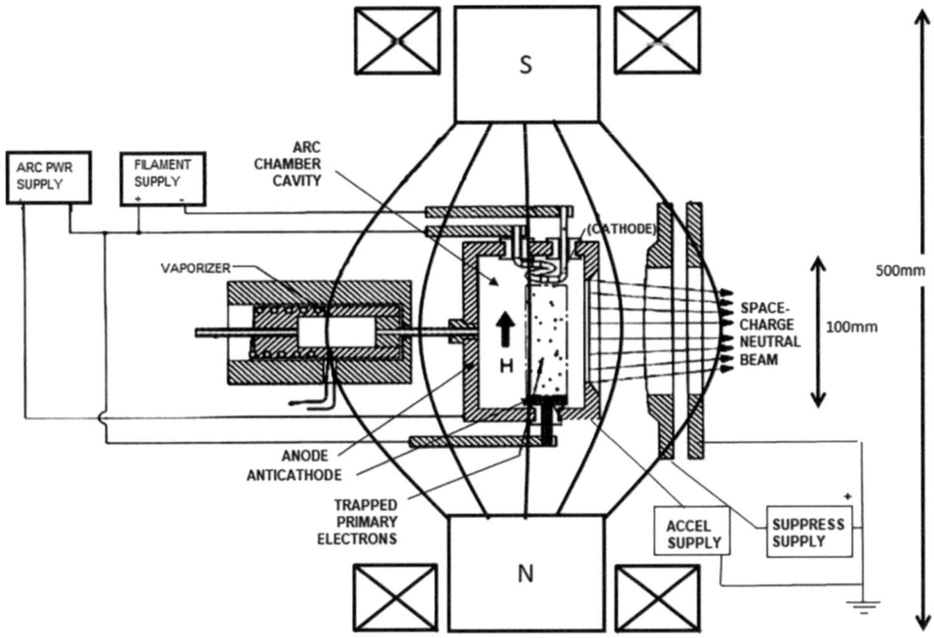

Figure 1.
A high current source using a Penning Ionization Gauge Trap (HC PIG ion source). Note the convergent ribbon beam formation, triode electrode arrangement with negative electron suppression electrode, and the non-uniformity of the magnetic field. The exit slot is 3 mm wide.

In commercial implanters it was found necessary to add two further features to most heavy-ion beamlines to achieve and preserve this neutralization: (1) a second negative electrode sandwiched between two grounded electrodes as a triode structure through which the beam passed, to isolate the bulk of the beam from any interactions with the beam target, which was often a silicon wafer covered in insulating photoresist, which could charge to a high potential, and (2) a source of low-energy electrons (such as a plasma flood gun) to suppress significant potentials from developing on insulating surfaces. In many systems, similar electron guns add electrons to the main beam, but these are unnecessary if the beam is fully surrounded by conductive grounded walls.

4. Beam hash and instability: the limit to analyzed beam current

It had been found early on in the Calutrons that there was a practical limit to the high beam currents that could be reliably analyzed, due to a phenomenon referred to as 'hash' on the beam—this was quasi-random high-frequency noise of several hundred kHz, which disrupted the space-charge neutralization of the ion beam, resulting in a loss of mass resolution and an inability to transmit higher currents [1, 2]. This phenomenon was not fully understood, but I had observed it when attempting to increase the transmitted current of arsenic beams above the specification. The beam could clearly be observed to blow up, and to exhibit hysteresis as certain source parameters were tuned. I later observed a similar phenomenon on a test stand I was working on in 1990 and 1991, where I had developed a PIG ion source delivering significantly more low-energy current than prior sources in a low-emittance ribbon beam, but when I tried to transmit it through an analyzing magnet, the beam blew up, even in the space before entering the analyzer magnet. The ribbon beam source I was using, like that in the NV10 implanter, had a slightly concave front aperture from which the beam was extracted, with the electrodes shaped to be parallel to this curve, so that the beam converged in the breadth dimension. The universally held belief (despite the Calutron experience) was that a magnetic field helped space-charge neutralization, but our observation was that the magnetic field could disrupt it. We could turn the magnet on and off, and watch the magnetic field cause the disruption of the beam exiting the ion source and entering the magnet. We made Langmuir probe measurements of electron temperature in the space very close to the beam, and we observed that applying the magnetic field to the beam could trigger a dramatic increase in electron temperature.

Analysis of the beam size, current density, perveance etc., and the application of some standard plasma science, led us to hypothesize that under certain conditions, unstable ion sound waves could be transmitted through the plasma within the ion beam. We never saw the beam blowup below a certain threshold current, but blowup would occur suddenly as the magnetic field was raised, only when certain quantifiable conditions were met, namely that the ion plasma frequency, Ω_{pi} exceeded the ion cyclotron frequency, Ω_c, by a significant factor, of the order of 2.

$$\Omega_c = \frac{qB}{M} \tag{3}$$

$$\Omega_{pi} = q\sqrt{\frac{n_i}{\varepsilon_0 M}} \tag{4}$$

The cyclotron frequency in ion implanters is typically in the range from 200 kHz to 1 MHz, and when beam blowup is observed it is accompanied by chaotic hash on the beam at frequencies above this threshold. Note, however, that n_i in Eq. (4), the density of ions, is not a well-defined quantity, as there are several ion populations present, whose interaction is not straightforward.

$$n_i = n_b + n_r \tag{5}$$

where n_b is the density of ions in the beam (not uniform) and n_r is the density of slow ions generated from the residual gas.

$$n_b = J_b\sqrt{\frac{M}{2qV}} \tag{6}$$

The slow ion density n_r is very difficult to evaluate, as it is proportional to the pressure, the beam current density, and the local potential distribution, since these positive ions will be weakly accelerated away from the beam center; the potential distribution is affected by the electron temperature, so n_r will drop as soon as the beam neutralization is perturbed. The pressure is usually related to the beam current and to geometric factors, and is rarely well-known in the center of a magnet.

$$n_r = \frac{I_b n_0 \sigma}{2(t_b + b_b)v_i q} \tag{7}$$

where t_b and b_b represent the beam thickness and breadth within the magnet, v_i is the velocity of the slow ions as they are repelled from the beam by its plasma potential, n_0 is the residual gas density, and σ is the total cross section for slow ion generation by the beam. To further complicate the picture, the beam ions contain multiple species, and Alexeff [6] showed that this can create additional instabilities. The beam ions are traveling at supersonic speed in the plasma. White and Gray Morgan presented a paper [7] in the form of a poster at the IEEE Plasma Science conference of 1991, in which scaling with the background gas ion density was demonstrated, and discussions with other participants including Igor Alexeff, founder of IEEE Plasma Science division, established that our scaling law was consistent with his models [6], and further that the geometry we were using could function in a similar manner to a klystron, and amplify the instabilities with self-feedback. When the ion density fell below the threshold we describe, the beam was stable.

5. The first DC ribbon-beam architecture

Moore's Law predicted rapid scaling up of the number of devices on a single chip, and this required several developments: higher currents at lower energies (as devices became smaller), larger silicon wafers, moving to 200 mm diameter in the 1980s, better control of beam angles, and a strong preference for serial wafer processing to avoid the inefficiencies of processing in fixed-size batches. Serial processing was used in most semiconductor manufacturing processes, but high-current implantation still processed wafers in batches, resulting in higher costs. Applied Materials had acquired Lintott Engineering Ltd. In 1980, and in '85 was manufacturing a somewhat successful batch high current implanter, the PI 9000, but as larger wafer sizes were considered, the future looked uncertain. Applied's other semiconductor tools all used serial processing.

The ASM 220 medium-current ion implanter used a horizontally scanned beam which was rendered parallel by means of a non-uniform magnet bending on average about 10 degrees [8, 9]. A single wafer was scanned mechanically through this beam in the vertical direction, to achieve uniform doping. This technique provided unprecedented angle control between beam and the wafer, excellent uniformity, and high efficiency (**Figure 2**).

In conjunction with marketing and technical colleagues at Applied Materials, White in 1989 proposed (a) the acquisition of the ASM implanter division, and (b) the development of a serial-process high-current implanter using a wide DC ribbon beam [10], with a similar mechanical wafer scan mechanism to the ASM 220; and the two implanters would be complementary for medium and high-dose applications, at least on 200 mm wafers. We anticipated that the next silicon wafer size would be 300 mm, and we sized our high-current proposal for a 330 mm wide beam. Applied Materials did not agree to these plans, and as a result, Diamond Semiconductor Group (DSG) was founded in 1991, with a mission of developing such an implanter.

In that intervening two-year interval, as described above, White and Gray Morgan had experimentally confirmed the scaling laws of the beam instability problem [7]. DSG concluded that the use of a convergent slit on the front of the ion source raised the ion density within the analyzing magnet above the threshold for beam instability at those beam currents and energies which the market desired. DSG's goals now included

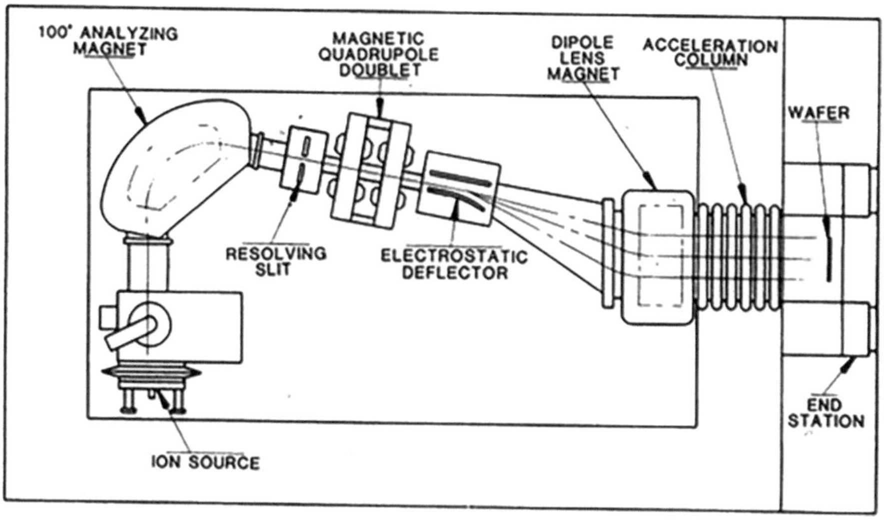

Figure 2.
The ASM 220 ion implanter in which a beam is scanned horizontally, then collimated by a dipole lens magnet, to generate a parallel uniform scanned beam.

maximizing the beam size within the analyzing magnet, to keep the ion density low enough that Ω_{pi} remained below Ω_c, at beam currents of 5 mA at 5 keV or less.

The resulting design was the SHC-80 ion implanter [11] from which the VIISta-80 and other descendants under the VIISta tradename have manufactured and sold by Varian, and after acquisition by Applied Materials are still being produced and sold. The key design decision was to use a horizontally oriented *convex* ribbon beam ion source, and to produce a horizontally divergent ribbon beam to enter an analyzing magnet and be sharply refocused at a resolving aperture. This was loosely based on the design of the Applied/Lintott PI9000 ion implanter (See pages 241 and 242 of [1]). Then this analyzed beam was allowed to diverge again and was shaped by a 70-degree magnet into a parallel beam greater than 300 mm in width. This beam then implanted single 300 mm wafers which were mechanically vertically translated through it (**Figure 3**) [12].

6. Ribbon beam flat-panel display implanters

Diamond Semiconductor Group then developed an implanter beamline with a ribbon beam width of >620 mm, for use in flat-panel display implanters made by Mitsui Zosen [13]. Subsequently White developed methods of changing the ion source orientation and achieving higher resolving power, by orienting the ribbon vertically, and bending it horizontally [14] with a magnet using 'bedstead' coils. This concept was used by Mitsui Zosen for producing a 1300 mm tall ribbon beam (**Figures 4** and **5**).

To operate an HC PIG ion source with a 150 mm long slot, and produce a uniform but expanding ion beam, the ion source magnet had to generate a substantially uniform field down the length of the ion source, the y-direction—yet the overall yoke height was only ~500 mm, as it was in **Figure 2**. Usually this creates a minimum field at the center, and substantially greater field near the poles. But here the poles were shaped to cause a rapid increase in a direction transverse to the beam direction and normal to its breadth, which I label x, which improves the uniformity along the

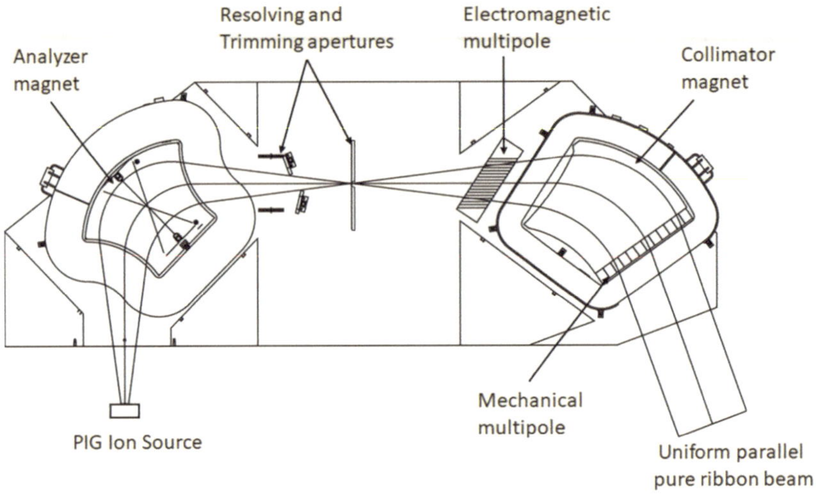

Figure 3.
The SHC-80 beamline. The divergent beam from the HC PIG ion source is refocused through a resolving aperture, diverges again and is rendered parallel by the 70 deg. collimator magnet.

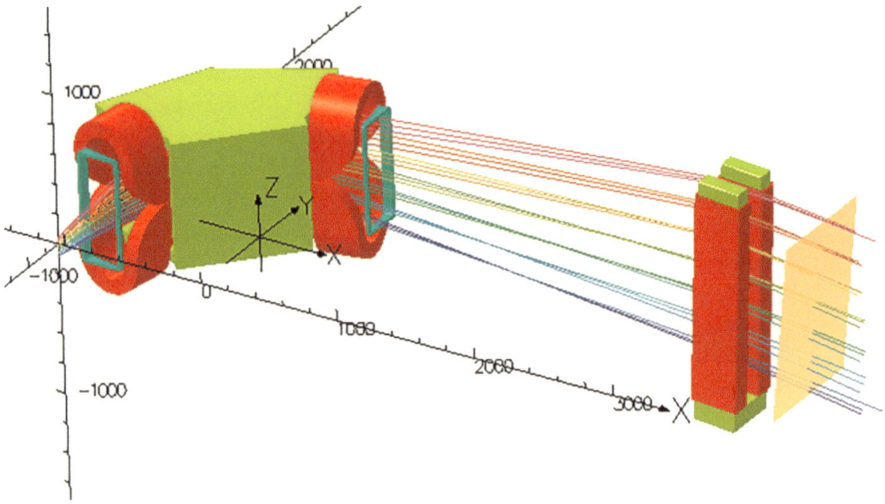

Figure 4.
A vertical ribbon beam 150 mm tall at the ion source expands vertically, is analyzed by a 90° magnet with 'bedstead' coils, expands to 1300 mm, and is collimated by a U-multipole [15], which also adjusts the uniformity.

central axis. In this manner we were able to achieve +/− 2% uniformity in the arc chamber and permit substantially uniform ionization.

7. The control of parallelism and uniformity in ribbon beams

The approach pioneered in the ASM220 implanter [9] (**Figure 2**) for collimating scanned beams was easily adapted for DC ribbon beams. Varian had acquired the ASM implant division and now licensed the technology of the SHC-80 implanter of DSG, so the first prototype of the SHC-80 used the ASM-220 (now E220) wafer handler. The

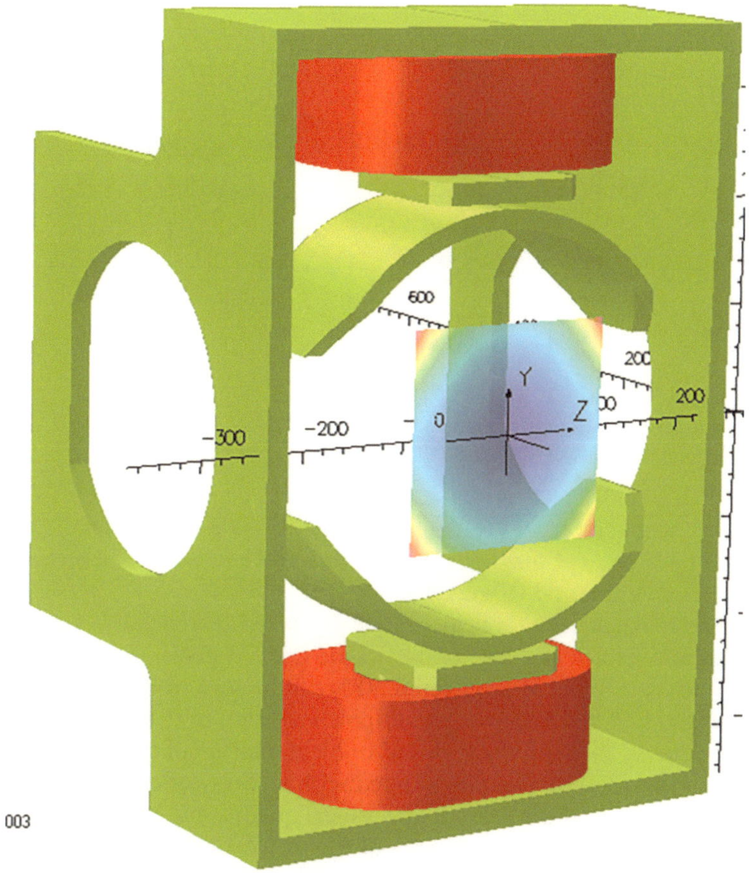

Figure 5.
The ion source magnet with optimized poles and yoke for a more uniform source magnetic field, as used with a 150 mm broad expanding beam in the implanter of ***Figure 4****.*

SHC-80 [12] design shown in **Figure 3** used a uniform field 70-degree magnet, which provided the correct amount of horizontal focusing to render the strongly diverging beam from the resolving aperture parallel, but also provided some vertical focusing, thereby delivering a substantially parallel beam in two dimensions to the endstation of the implanter.

The control of uniformity was a more difficult problem. The first approach was crude and simple: to trim those regions of the ribbon beam where the current was too high with mechanically adjustable trimmers, nicknamed 'magic fingers'. To minimize sputter contamination of the silicon, these fingers were located inside the first magnet, at the local maximum in the beam width. Unfortunately, it was found that these only rendered the beam less uniform, for as each finger neared the beam, a plasma sheath formed beside it and the electric field in the sheath deflected the beam ions that came close to it sufficiently to generate peaks and valleys on the current profile across the final beam.

However, this provided a hint for the successful development and use of linear multipole lenses to correct the uniformity [16]. To raise the linear current density (measured in mA/cm along the beam breadth dimension) it would be necessary to slightly deflect ions from neighboring regions toward the low-density zone. To lower

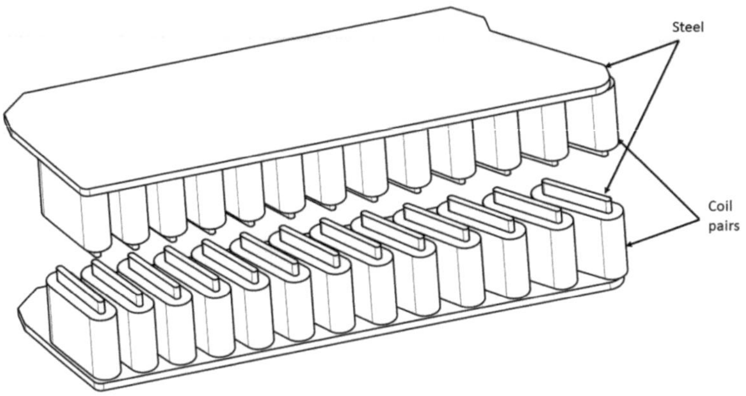

Figure 6.
The multipole for uniformity control from the SHC-80 implanter.

it, the opposite would be true. We therefore used linear multipole lenses, being a horizontal array of pairs of magnetic poles above and below the ribbon beam, to successfully achieve this end (**Figure 6**).

An alternative mechanical multipole style consisted of a movable set of fingers made of magnetic steel at the edge of one of the magnet poles, thereby allowing the local field to be slightly raised or lowered. This can be seen in **Figure 3**.

Subsequently, White developed a multipole variation of the 'Piccioni' quadrupole lens [17], to simplify the manufacture and implementation of these multipoles, comprising two parallel steel bars, each wrapped with an array of coils. Energizing one pair of opposite coils on these bars creates a local quadrupole field component, which raises or lowers the linear current density of the region of the ribbon beam passing between that pair of coils. This arrangement produces a smoother control of the uniformity from a simpler structure, though the underlying theory is the same. Passing a similar current through all the coils in this multipole generates a uniform broad quadrupole field, which can be used for correcting the beam parallelism. A pair of such quadrupoles can be used to simultaneously fine tune the breadth, the parallelism and the uniformity of a ribbon beam (**Figure 7**).

A further variation of this was developed by White [15], in which one multipole has the two steel bars linked by a piece of magnetic steel at one end, creating a

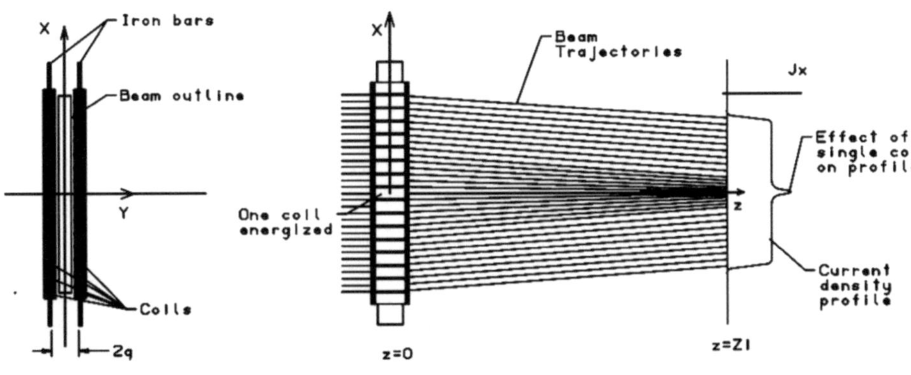

Figure 7.
A multipole lens based on the Piccioni quadrupole lens concept.

U-shaped magnetic steel yoke. This version can develop an overall dipole field superimposed on quadrupole and multipole field components, and it can be useful for introducing an overall deflection of 5–15 degrees into the beam. This can be seen in **Figure 4**, as used in Mitsui Zosen flat-panel display implanters.

8. Intense broad ribbon beams without local dipole magnets

8.1 Background

During and after the 1990s, almost all companies used ion sources based on the Bernas source with enhancements by White of a passive anticathode electrode at the opposite end of the arc chamber from the hot filament cathode, both being identically biased at about −60 to −120 V. This was later improved by Horsky [18], who replaced the hot filament with an indirectly heated cathode, (IHC) a block of tungsten with a hollow interior, heated by electrons from an internal hot filament, biased several hundred volts negative. This has a much greater service life, typically 500 hours, than a White source (~150 hours) or a Freeman source (~60 hours). I refer to all these variants as HC PIG (Penning Ionization Gauge) discharge sources. Early reports attributed little benefit to this HC PIG arrangement, but White and Westner found that when running BF_3 gas, the fast electron density could be significantly higher, and since B^+ ions are generated from BF_2^+ ionic molecules, which in turn are produced from BF_3 gas, the higher electron density results in a much higher fraction of B^+.

In an HC PIG source, the locally applied magnetic field confines the fast primary electrons from the hot cathode radially, so they cannot reach the chamber walls at anode potential, and the cathode/anticathode electrodes repel the electrons from the ends. The result is the efficient creation of Penning Trap, in which fast electrons with energies between about 75 and 140 eV, depending on the cathode voltage, are confined. These efficiently ionize gas or vapor molecules within this zone.

The cross section for elastic scattering rises dramatically as the energy of electrons is reduced; as a result, those electrons which excite or ionize the gas, losing significant kinetic energy, then interact more strongly with each other through elastic collisions, and form a relatively cool plasma. The cooler electrons can diffuse quite readily across the magnetic field—it is only effective at confining the faster electrons, and so a cool plasma forms in thermal equilibrium at a temperature of 4–8 eV, and following classical plasma laws reaches an equilibrium voltage positive with respect to the potential of the anode walls (**Figure 8**).

Now consider the problems of developing much larger ribbon beams, with sizes from 450 mm to say 2 m. With a much longer ion source, much higher beam currents could be obtained. The linear current density at the ion source which HC PIG sources produce is over 10 mA per cm, but they cannot be scaled up in their present form. The 400 mm Calutron sources were immersed in a gigantic uniform magnetic field; this approach is possible but completely uneconomic.

8.2 Design goals

We require an ion source producing a uniform ribbon beam with a breadth unlimited by practical scaling laws, at least 1.5 m, but with the thickness of the beam from prior sources, typically 3 mm but sometimes about 5 mm, with no magnetic field component along the axis of the arc chamber. This gives an aspect ratio of 360–500,

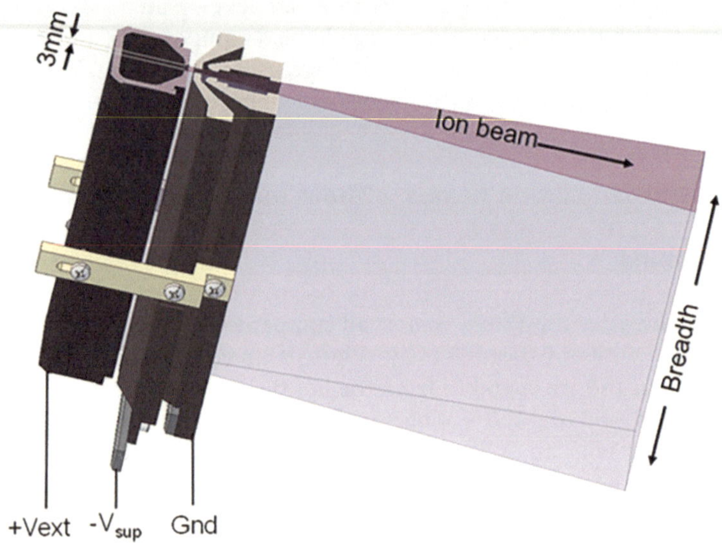

Figure 8.
A long arc chamber with a 3 mm slot can generate a broad ribbon beam; the uniformity and efficiency of ionization along the arc chamber is crucial.

and by Eq. (2) this gives very high upper limits to beam current, well over 1 amp of ions at keV energies. We require the ion source to be efficient, which means trapping 60–140 eV electrons in a tight efficient zone behind the extraction slit, and then introducing a uniform distribution of gas or vapor to be ionized.

We also want an analyzing device to separate ion species with modest resolving power, useful for a number of light ions, which also has a breadth essentially unlimited by scaling laws.

8.3 The CusPIG confinement trap

One of our first thoughts was to use a multicusp arrangement, the most obvious simple arrangement being a quadrupole magnetic field, oriented transverse to the beam breadth dimension, and extending uniformly down the arc chamber. However, we knew from prior work with multicusp trapping that it was far less efficient than a Penning trap, and fast electron densities in multicusp confinement were significantly lower than in a Penning trap, because of significant electron losses occurring at the cusps. So we looked for a way to block the four magnetic cusps with electrodes at cathode potential. An initial success with this arrangement led to a simplification: make the chamber walls at cathode potential, and place the anodes in a location where the magnetic shielding would be most efficient:

The arc chamber was an almost square section extruded in our first prototype to a length of 350 mm, but this could have been longer without any practical limit. At one end is an indirectly heated cathode, and at the other end is a wall, all the walls being at cathode potential. Magnetic poles run along the length of the arc chamber. The version shown in **Figure 9** uses permanent magnets, but during the development, we used electromagnets. In principle there are four poles, but one pole is cut away for the beam exit slit. This producs a distorted quadrupole magnetic field extending along the length of the arc chamber, with the field components entirely in the plane of the cross

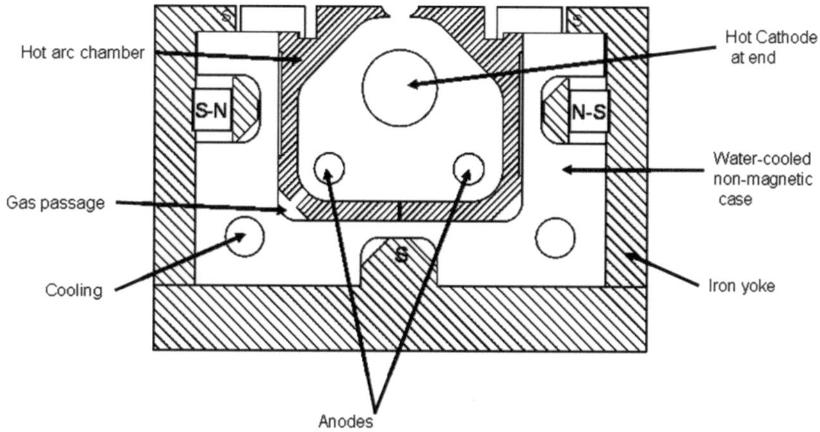

Figure 9.
Cross section of the CusPIG arc chamber, permanent magnet version. The magnetic field has a null running the length of the arc chamber, in line with the hot cathode.

section shown. There is a field null located within the arc chamber, just behind the exit slit. Two rods of molybdenum run the length of the arc chamber and are biased positive at a voltage between about 60 and 150 V. These are the only surfaces at anode potential, and are located where fast electrons from the hot cathode cannot reach them without crossing magnetic field lines with an integrated strength of about 0.3 T. mm. The crucial feature is that the magnetic cusps do not leak electrons, because they are blocked by walls at cathode potential. This is a form of quadrupole Penning trap, where cathode potential is placed at the magnetic cusps, and anode potential only exists in the region away from the cusps. We coined the name CusPIG to describe this geometry [19].

There are other alternative implementations: the quadrupole magnetic field could be re-oriented by rotating 45 degrees, the magnetic poles being located near the corners. But the basic principle is a quadrupole magnetic field enhanced by Penning Trap potentials.

It was found that the efficiency of confinement varied very little with the strength of the field, so this permanent magnet arrangement became standard. The simplest construction provides two extended north poles each excited by an array of ¼" Neodymium-iron-boron magnets, one at each side, extending the length of the arc chamber, i.e. for the breadth of the beam. The arc chamber was graphite, and allowed to run hot, but was surrounded by a water-cooled aluminum case. A magnetic yoke surrounds three and a half sides of the arc chamber, provides a discrete south pole at the base, and provides two rudimentary weak south poles on either side of the beam exit slot. This creates a quadrupole field with a significant sextupole component; the fields are strongest in the back of the arc chamber and weaker near the exit, while the null line is located just behind the exit slot. Anode potential is confined to the two metal rods located where the magnetic field is strongest. This is the preferred construction.

Within a plasma magnetized by such a quadrupole field, fast electrons have a cycloidal path around magnetic field lines with a constantly changing curvature. Unless scattered by an ionizing event, the fast electrons have very low mobility normal to the field lines. But there will be a weak electric field normal to the field lines, giving rise to some electron cross-field drift at velocity $\mathbf{v} = \mathbf{ExB}$ along the length of the arc chamber, in opposite directions in adjacent quadrants. Modeling also appears to reveal chaotic drifting along the arc chamber. Electrons are magnetically blocked from reaching the anodes, and are electrostatically blocked from reaching the walls. On reaching the end of the arc chamber there is a high chance of the electron hopping into another quadrant and returning in the opposite direction. By means of modeling the ballistic trajectories in the modeled magnetic field, and approximated electric field, using OPERA/TOSCA, we show that the whole arc chamber fills rapidly and uniformly with fast electrons except very close to the walls.

In **Figure 10**, the left view shows a transverse cross section of the electron trajectories as produced by the source of **Figure 9**. The Penning trap arrangement prevents any electron loss where the magnetic confinement would fail. The color represents the electron velocity; blue is low velocity, red is high—so the deceleration and reflection of electrons at the cusps can be seen. The quadrupole magnetic confinement shields the anode rods, while providing a zero field zone near the center of the chamber. Weak electric fields in and near the plasma sheath create **ExB** drift of the electrons, weakly emulating a magnetron racetrack, to help circulate the electrons around the discharge zone and enhance uniformity. The electron motion is very complex, consisting of cycloidal motion, but since the energy changes rapidly in the plasma sheath, this motion is somewhat chaotic. It extends rapidly down the length of the arc chamber, as shown.

Gas is introduced through orifices shown in **Figure 9**, blocked by the anode rods from direct line-of-sight to the exit slot. This makes for very efficient ionization of the introduced gas. If the mass flow is precisely known, and the ion current is precisely known, one can calculate the ionization efficiency, in terms of the fraction of the gas that is introduced which exits as beam ions. This is a function of many factors, the arc voltage and current, which determine both the fast electron density and the cross section. Operating the ion source for a constant ion current, as the gas flow is reduced, a higher arc current is required. Using ~6A of arc current and an anode voltage of 120 V, with argon gas, the ionization efficiency had a highest value of ~30% at 8.9 sccm.

The cathode is at one end of the arc chamber. The density of ionizing electrons is observed to fall with distance from the cathode, since electrons which ionize an atom or molecule lose energy, are less effective, and rapidly diffuse out of the core zone. We found that this attenuation was a strong function of the anode voltage—the higher the voltage, the further the mean distance that the electrons traveled before being lost from the useful population. We have measured the attenuation of the ionizing electrons as a function of distance from the cathode by measuring the uniformity of the ion beam. The result is informative and useful. First, an arc voltage minimum of about 38 V is required to get a discharge, and the current density falls very rapidly with distance. At 70 V, the electrons travel 3 m before the ionization falls to 1/e of the initial value. (Incidentally, because their paths are not straight, this is equivalent to about

Figure 10.
At left is a ray-trace of multiple electrons, viewed from the end of the arc chamber. At right is a side view of the trajectory of a single electron.

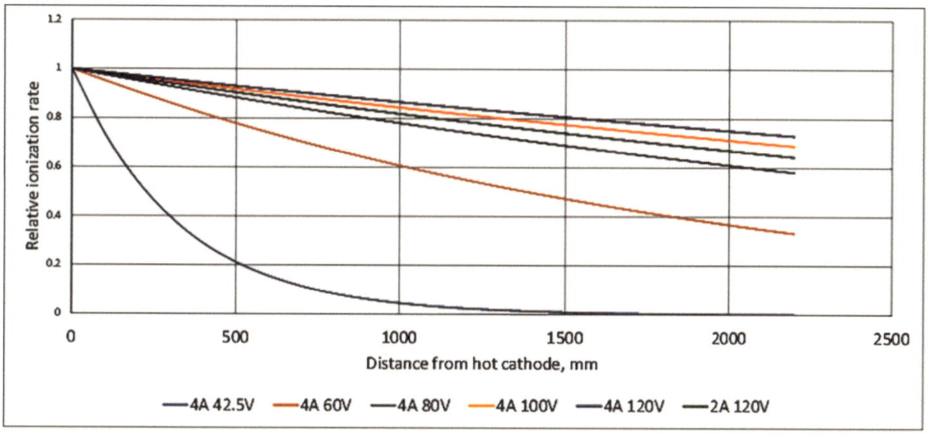

Figure 11.
The measured drop in local ionization rate as a function of distance from the cathode.

20 m of actual travel). At 120 V the attenuation was 20% at a distance of around 1.5 m from the cathode. It should be pointed out that because of the cycloidal electron motion, the true mean free path is many times greater than this. But as a practical matter, with 100–120 V anode voltage, the non-uniformity would be about 20% for a 1500 mm broad beam, and if a cathode were placed at each end (which is easily done), the intrinsic non-uniformity would fall to well under 1% (**Figure 11**).

The spacing and profiles of the electrodes, and the suppression voltage, are modeled with OPERA/ELECTRA, and can be optimized for voltages from ~2 keV to 60 keV with ease. **Figure 12** shows a 3 keV case. The beam is self space-charge neutralized, without the addition of neutralizing electrons from a gun. Great care was taken to avoid any unshielded connections to electrodes, etc., which should disrupt this neutralization. For the present we do not use a dynamically adjustable electrode gap, because the high precision required to optimize tuning along the full arc chamber length would be very mechanically challenging.

Ionizing collisions are not the only inelastic collisions that can occur. Neutral excitation and resulting subsequent light emission also occur. For this reason, viewing the glow of the ion beam can be useful but misleading, for the glow is generated by electron transitions resulting from charge-exchange and other collisions with residual gas, so the emitted light is mainly from excited neutral atoms, so some of the halo around the beam may come from excited neutrals which travel outside the beam before decaying.

Discharge uniformity can be improved by two methods. The first is simple to implement. The gas flow to the different gas ports can be adjusted in real time to adjust the uniformity of the extracted beam. Crude adjustment is particularly easy: the current on the suppression electrode is proportional to ions which strike the electrode—ideally zero. The suppression current therefore measures beam de-tuning. Initially the

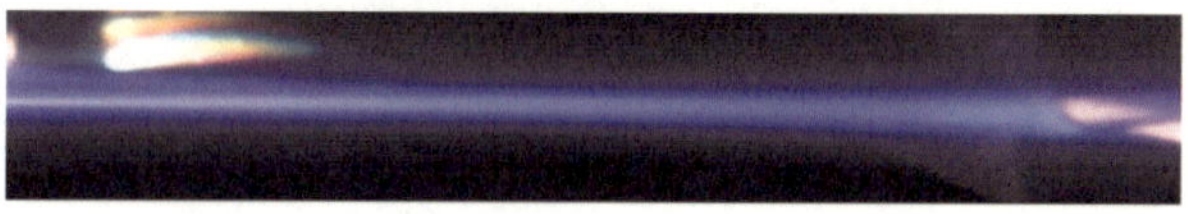

Figure 12.
Photograph from above a 350 mm broad vertical argon ribbon beam of 120 mA at 3 keV. The beam at left is emerging from the third electrode of the triode. Divergence is ~+/− 20 mrad. Scale is about 50%.

suppression voltage can be adjusted to minimize this current, but then differential gas flow adjustments (made with the total beam current or arc current held constant) can be made to further reduce the unwanted suppression current, which is to effectively make the beam divergence—and hence its uniformity—more uniform. A second method, as discussed above, is to place a hot cathode at each end. If the ion source is well over 2 m long, it may be beneficial to add an additional hot cathode at a central location.

9. Extendable mass-Analyzer for meter-scale ribbon beams

Conventional magnetic mass analysis of ribbon beams broader than ~500 mm is difficult and does not scale well. The gap between the poles of a conventional dipole magnet is much smaller than the bending radius of the trajectories in the magnet, to avoid enormous aberrations, stray fields and related difficulties. As the beam breadth is increased, the cost, weight, power of the magnet all increase more steeply than with the square of the beam breadth. The fringe field grows in proportion to the pole gap and is a source of aberrations and crosstalk with other system components, which also increase in magnitude with the square of the pole gap.

A number of partial solutions have been developed, as described above, but these have been incremental improvements with significant tradeoffs. Aitken proposed a radically new analyzer in 2002 [20] which used a new approach, from which the present design has taken inspiration; however, it was more complex and larger than the present device for given beam parameters. I attempted a design using an E-shaped magnetic yoke [21], which was successfully tested, but the present design is simpler and lighter, with better resolution. In a conventional magnet the focal length scales with the bending radius, and the bending radius of necessity scales with the gap between the poles, so for a beam 2 m broad, one can expect the focal length to be several meters, while the aberrations would be severe, and the weight to exceed a hundred tons.

The magnet design presented here addresses these issues, and differs from conventional dipole magnet configurations in the following ways:

- Ampere-turns are independent of beam breadth.
- Aberrations are independent of beam breadth.
- Overall dimensions and weight are a fraction of conventional magnets
- System path length does not increase with beam breadth
- Resolving power is adequate for boron and phosphorus, but not for medical isotope separation
- Beam current limit is proportional to beam breadth.

A system meeting these goals can have no component of magnetic field in the beam breadth direction, or it must violate these requirements. But if a field component normal to the breadth deflects the beam, then after a finite short distance the opposite field restores the beam direction, there must be a component of field in the gap between these zones in which a component of field is orthogonal to the s-shaped beam trajectories. Thus the solution is a magnet which deflects initially in the breadth direction, and produces a 3D bent S-shaped set of trajectories, hence the acronym U3DS [22].

9.1 The U3DS design

This magnet comprises a single U-shaped iron yoke as shown in **Figures 13** and **14**, providing two rectangular zones of uniform dipole field extending across the large dimension of the ribbon beam. Three optional iron bars labeled P1b, P2b, and P3 are shown placed on the opposite side of the beam. They are not mandatory, but serve the auxiliary functions of homogenizing the field, lowering the ampere-turn requirement, and thus the power requirement, and enabling some reduction in aberrations. Two simple identical rectangular coils provide the magnetic induction.

The U3DS magnet deflects the beam in an S-shaped path, bending in its major (breadth) x-direction and back, as illustrated in **Figure 15**, causing an offset in its path, and this deflection increases with the field strength. The net angular deflection in the breadth dimension is zero and is achromatic. However, unexpectedly, the beam is also deflected away from the exciting coils through an angle which is proportional to

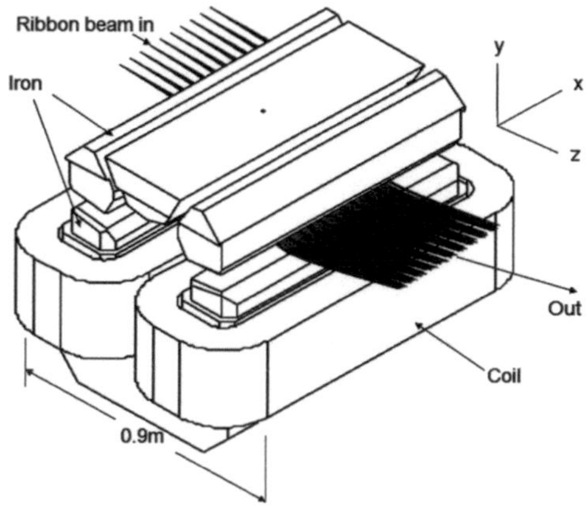

Figure 13.
Perspective view.

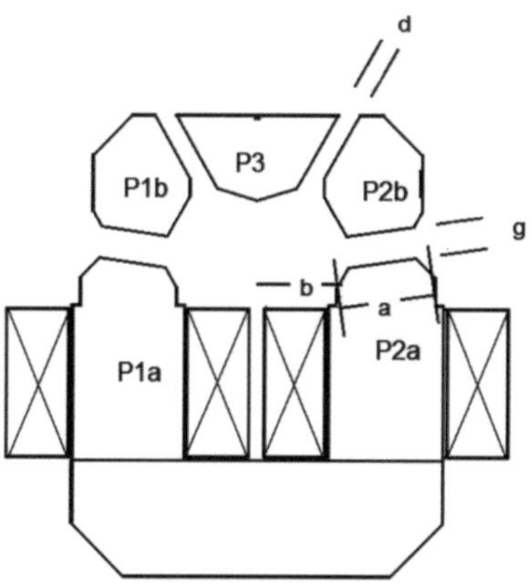

Figure 14.
Sectional view of the U3DS analyzer.

the square of the field strength, and is simultaneously strongly focused in this direction. The trajectories have 3D S-shaped paths, and this y-direction deflection allows the device to be used as a spectrometer in the conventional manner. The dispersion is surprisingly high, and therefore so is the resolving power. The ribbon beam is bent through a modest angle of 15–25 degrees in its thin direction, and strongly focused, while the peak deflection in the S-shaped path in the breadth direction is about twice this amount. The dispersion achieved is twice that of a simple dipole with the same bend angle, so the performance is as good as a magnet bending ~50 degrees. The focal length is substantially shorter.

The new magnet provides unperturbed uniformity along the beam breadth direction because there is no intrinsic variation of any field component along the breadth, and this avoids several of the worst aberrations of conventional dipole magnets. Nevertheless, there is a first-order aberration limiting resolving power, proportional to the intrinsic random divergence spread in the beam breadth direction. This limits the resolving power in practical situations to about 30.

Manufacture of IG6 flat panel displays requires a beam broader than 1.5 m, able to analyze an 80 keV P^+ beam. A suitable U3DS magnet would weigh about 7 tons. I estimate that a suitable dipole magnet with a 1600 mm pole gap would weight about 100 tons. The beam current carrying capacity is many times greater, as discussed in detail below.

In **Figure 15**, three zones of magnetic field, q, r, and s, are labeled. Zone q is where the trajectory first passes between two poles (labeled pole P1a and P1b in **Figure 14**), where the field will be substantially uniform and orthogonal to the trajectory. The field will fall at the edges; the effective length of uniform field is denoted by dimension a in **Figure 14**, and the gap between poles by g. The magnetic field direction is reversed in the right-hand half. Magnetic field zone s in **Figure 15** corresponds to this reversed reflection of zone q, adjacent to pole P2a. Central zone r is characterized by a predominant component of field from left to right in both views, but this field is non-uniform, and the field lines have some curvature. Poles P1b, P2b, and P3 are

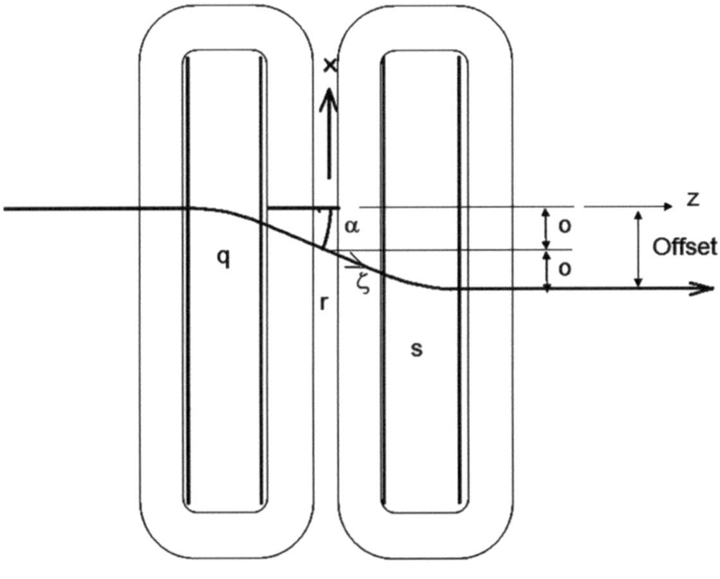

Figure 15.
Plan view: S-shaped path of reference trajectory.

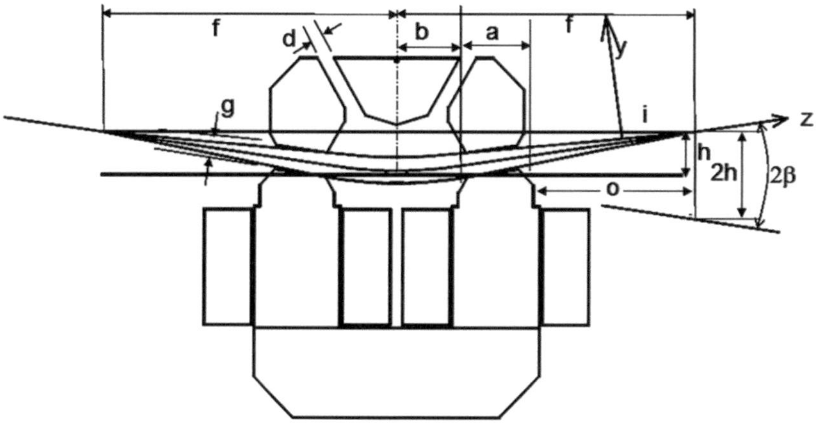

Figure 16.
Section with trajectories and optics.

supplementary pole pieces which concentrate the H-field where it is required, thereby greatly reducing the required coil current and power. They can be used to control the field shape to minimize aberrations.

The deflection in the y-direction arises because in the central zone r the ions have a large component of motion in the x-direction, and the magnetic field is predominantly in the z-direction—hence the deflection is normal to both—and is in the y-direction. Being proportional both to the z-field and to the amount of motion in the x-direction, which is itself a function of B, the amount of y-deflection scales with B^2.

Optical elements are defined as illustrated in **Figure 16**. The focal length can be estimated from the geometry, given that the total deflection 2β = ~2 h/f and 2 h = ~3 g, and assume g = d, using symbols defined in **Figure 16** This works well in practice. Greater accuracy requires full 3-dimensional trajectory modeling.

The beam plasma instability in magnetic fields discussed above applies to any magnet design. Experimentally, as discussed above, the background ion plasma density seems to be the most important factor. Ion sound waves in this plasma can carry instabilities through the magnet and cause them to be amplified, if the threshold for this instability is significantly exceeded. For a rectangular beam of high aspect ratio A > > 1 whose width (narrow dimension) within the magnetic field B is t_b, and in an ambient pressure P, the dimensionless figure of demerit for instability in residual gas is.

$$\Omega_{pi}/\Omega_c \cong \frac{M_i}{B}\sqrt{\frac{J_b t_b P\sigma}{\varepsilon_0 q k T_0 \sqrt{M_i k T_e}}} \tag{8}$$

M_i is the ion mass, B is the magnetic field in the beam, J_b the current density, t_b is the beam thickness in the magnet in the narrow direction, and thus the quantity $J_b t_b$ is the one-dimensional current density along the beam breadth direction, a quantity of direct interest. σ is the cross-section for slow ion production by charge exchange with beam ions, T_0 is the residual gas temperature, and T_e is the plasma electron temperature. This equation follows from Eq. (7), assuming a very broad beam whose thickness can be neglected in this context. As an example, for the magnet used as a worked example above, the figure of demerit for the background gas instability for a beam of

15 keV P^+ at 0.43 T Tesla, in a pressure of 2×10^{-5} mbar, at a current of 500 mA per meter, is ~1. This is a current well in excess of existing commercial requirements; the instability usually becomes problematic when this ratio is ~2 or more; so for practical purposes we do not expect to see beam instability with this device.

10. Conclusions

The CusPIG ion source can generate current densities of ~30 mA/sq. cm of many ion species, because of its efficient trapping of ~100 eV primary electrons, and has a slot exit ~3 mm wide, producing a ribbon beam, whose breadth has been demonstrated to 350 mm and can be extended to meter-scale dimensions; thus the linear current density has a maximum of around 1ampere per meter breadth.

The U3DS analyzer design can analyze meter-scale broad beams, while being less susceptible to the 'hash' instabilities of high-current beams in magnets. It allows the maximum current to increase in proportion to beam breadth. Paired with the CusPIG ion source, the U3DS magnet can analyze and transmit current densities approaching 1 ampere per meter, at an energy of 15 keV, from a 3 mm slit. With such an ion source it can achieve a modest resolving power of around 30.

Author details

Nicholas R. White
Albion Systems, Manchester, Massachusetts, USA

*Address all correspondence to: nick.white@ieee.org

References

[1] Freeman JH. Canal rays to ion implantation, 1886–1986. Radiation Effects. 1986;**100**:161-248

[2] Yergey AL, Yergey AK. Preparative scale mass spectrometry: A brief history of the Calutron. Journal of the American Society for Mass Spectrometry. 1997; **8**(9):943-953

[3] Bernas R, Kaluszyner L, Druaux J. Sur la neutralisation de la charge d'espace des faisceaux d'ions positifs par accumulation d'électrons; étude du temps de neutralisation. Journal de Physique et le Radium. 1954;**15**:273

[4] Rose P. A history of commercial implantation. Nuclear Instruction and Method. 1985;**B6**:1-8

[5] Ito H. High Quality High Current Generation and Transport Systems Including Plasma-Based Space-Charge Neutralizer. UK: University of Salford; 2003

[6] Alexeff I. Instability threshold for a Calutron (isotope separator) with only one species, IEEE trans. Plasma Science. 1983;**PS-11**(2):90-91

[7] White NR, Grey Morgan T. IEEE Intl. Conf. On Plasma Science. Williamsburg, VA, USA; 1991. Available from: www.Researchgate.net

[8] Berrian DW, Kaim RE, Vanderpot JW, Westendorp JFM. Nuclear Instruments and Methods. 1989;**B37**:500

[9] Kaim RE, van der Meulen PFH. Nuclear Instruments and Methods. 1991; **B55**:453-456

[10] White NR. US. Patent 5,126,575

[11] White NR, Sieradzki M, Renau A. The ion beam optics of a single-wafer high-current ion implanter. 1996 Intl. Conf. Ion Implantation Technology. IEEE; 1997:396-399

[12] White NR, Sieradzki M, Renau A. US Patent 5,350,926 (DSG 1)

[13] Satoh S, Degawa T, Watanabe H, Ujihara K, Oguro K, Shimamura K, et al. 1998 Intl. Conf. Ion Implantation Technology. IEEE; 1999:354-357

[14] White NR, Chen J. US Patent 7,112,789

[15] White NR US Patent 8,035,087

[16] White NR et al. The control of uniformity in parallel ribbon ion beams up to 24 inches in size. In: Duggan JL, Morgan IL, editors. Applications of Accelerators in Research and Industry. AIP; 1999:830-835

[17] White NR US Patent 7,078,713

[18] Horsky T. Review of Scientific Instruments. 1998;**69**:1688

[19] White NR, Westner AO. CUSPIG Ion Source for Large and Very Large Ribbon Ion Beam Systems. In: 21st International Conference on Ion Implantation Technology (IIT). Tainan, Taiwan: IEEE; 2016. pp. 116-119. DOI: 10.1109/IIT.2016.7882857

[20] Aitken D, IIT 2002, International Conference on Ion Implantation Technology (IIT), IEEE, Taos, NM, USA, pp. 448-451.

[21] White NR. US Patent 8,921 14th, 802

[22] White NR. U3DS, a compact mass-analyzer for large high current-density ribbon ion beams. In: 21st International Conference on Ion Implantation Technology (IIT). Tainan, Taiwan; 2016. pp. 1-4. DOI: 10.1109/IIT.2016.7882914